# DISTILLATION
# CONTROL,
# OPTIMIZATION,
# AND TUNING

## Fundamentals and Strategies

# DISTILLATION CONTROL, OPTIMIZATION, AND TUNING

## Fundamentals and Strategies

## LANNY ROBBINS

**CRC Press**
Taylor & Francis Group
Boca Raton   London   New York

CRC Press is an imprint of the
Taylor & Francis Group, an **informa** business

CRC Press
Taylor & Francis Group
6000 Broken Sound Parkway NW, Suite 300
Boca Raton, FL 33487-2742

First issued in paperback 2017

© 2011 by Taylor and Francis Group, LLC
CRC Press is an imprint of Taylor & Francis Group, an Informa business

No claim to original U.S. Government works

ISBN 13: 978-1-138-07382-1 (pbk)
ISBN 13: 978-1-4398-5748-9 (hbk)

| Library of Congress Cataloging-in-Publication Data |
| --- |

Robbins, Lanny.
    Distillation control, optimization, and tuning : fundamentals and strategies / Lanny Robbins.
        p. cm.
    Includes bibliographical references and index.
    ISBN 978-1-4398-5748-9 (hardback)
    1. Distillation. 2. Separation (Technology) I. Title.

TP156.D5R625 2011
660′.28425--dc22                                                          2010048806

**Visit the Taylor & Francis Web site at**
**http://www.taylorandfrancis.com**

**and the CRC Press Web site at**
**http://www.crcpress.com**

# Contents

# The Author

**Dr. Lanny Robbins**, president of Larco Technologies LLC, is a consultant. He retired with the title of Research Fellow after 37 years at The Dow Chemical Company at Midland, Michigan. In 2006 Dr. Robbins was elected to the National Academy of Engineering. He also received the Marston Medal, which is the top honor awarded by the Department of Engineering, Iowa State University. His expertise is in fundamental chemical engineering research and pilot plant process development, especially for separation and purification processes. He has developed unique technology for scale-down and scientific study of plant processes to generate experimental data for successful scale-up. He is the author of the liquid–liquid extraction chapters in *Perry's Chemical Engineer's Handbook*, sixth and seventh editions, and *Schweitzer's Handbook of Separation Techniques for Chemical Engineers*, first through third editions. Robbins has researched, developed, and implemented many separation and purification unit operations used at Dow.

Dr. Robbins is an inventor of two new liquid distributor designs for packed distillation, absorption, and stripping towers that are used in over 450 towers. Robbins developed a number of liquid–liquid extraction processes for product recovery and for process water purification. He developed the AquaDetox aqueous purification device and design technology for stripping residual solvents and impurities from water to the parts-per-billion range. He also developed other commercial wastewater purification technology to reduce trace impurities to unprecedented low levels of parts-per-quadrillion. Robbins developed the Sorbathene pressure swing adsorption process to remove hydrocarbons, solvents, and monomers from vent emissions. Robbins is the author of 19 publications, 18 U.S. patents, and more than 185 technical reports for Dow.

Dr. Robbins has served as an adjunct professor at Virginia Polytechnic Institute and Michigan State University. Robbins received the prestigious H. H. Dow Gold Medal from the board of directors of The Dow Chemical Company in 1993, and in 2003 he received the inaugural award for the Process Development Division of the national American Institute of Chemical Engineers (AIChE). Dr. Robbins received his B.S., M.S., and Ph.D. degrees in chemical engineering from Iowa State University.

## Chapter 1

# Introduction and Overview

Welcome to the *Strategy of Distillation Control, Optimizing Quality Performance, and Tuning Control Loops*. This first chapter is a short summary of the topics covered in this course.

## Learning Objectives

When you have completed this chapter, you should be able to

1. Understand the general organization of the content in the course
2. Know the course objectives
3. Know how to proceed through the course

## 1.1  Course Coverage

This book focuses on the fundamentals of process control of distillation columns. It covers the following topics:

1. The process variables for continuous binary distillation columns and four basic control strategies
2. The distillate and bottoms product quality performance objectives
3. The tuning of process control loops

When you finish this course, you will understand the fundamental separation and purification concepts to be achieved by a distillation column and the functional criteria that are critical for successful implementation of process control. Concepts for measuring and optimizing product quality performance will also be understood. You will also learn how process control loops for distillation columns can be tuned for stable operation with a balance between minimum variability from setpoint changes and excellent response to load disturbances. By approaching the subject this way, you will gain a fundamental understanding that can guide decisions for the design, operation, and troubleshooting of distillation process control systems.

This book is written from a perspective that was developed as the result of 37 years of industrial research at The Dow Chemical Company that focused on inventing, developing, and implementing industrial separation and purification processes. The work process for research and development is the scientific method, which will now be described.

## The Scientific Method

1. Define the problem and the opportunity.
2. Search and understand the state of the art.
3. Develop a hypothesis, that is, a concept or a model.
4. Design and run experiments to test the validity of the hypothesis.
5. Evaluate, summarize, and document the results.

The main problem in distillation process control is to separate and purify chemical components in liquid and vapor streams while shedding the disturbances that are imposed on the distillation column. The opportunity is to separate the components from a feed stream into new vapor and liquid streams that have increased economic value at a cost that is competitive with other producers. Conducting research in a large corporation provides the opportunity to apply the results of improved performance to many distillation towers in many different businesses.

Distillation is a mature technology that is well developed and tested by the scientific method. Two comprehensive books have been written by Kister[1,2] on the design and operation of distillation columns. The control of distillation columns has been widely studied for many years, many papers have been written, and several books have been published on the subject such as the one by Buckley, Luyben, and Shunta.[3] There is a large body of information to search and understand in the state of the art of distillation and process control.

During the last 20 years, there has been a continuing emphasis on improved product quality, performance, and reduced operating costs such as energy consumption, lost product, rework, maintenance, and labor cost per unit of product produced. Computer hardware and computer programs have been improved dramatically for simulating and modeling the distillation process and the dynamic response for the hypothesis in Step 3 in the scientific method. The chemical engineering fundamentals and mathematical hypotheses used to describe the design and performance of distillation columns are continually tested for their validity by plant operations in industry. Ultimately, the results are judged by the business that is responsible for the quality and profitability of the products produced by distillation and by the manufacturing department responsible for operating the process equipment.

## 1.2 Purpose

The purpose of this book is to present the fundamentals of process control of a distillation column as a separation and purification unit operation. This includes the critical concepts and functional criteria for the design, operation, and trouble-shooting of distillation process control plus the concepts of measuring and improving product quality performance. There is a prevailing need to strike a balance between understanding the concepts that are critical to exercising good engineering judgment and understanding the intricate details of each hypothesis. The focus of this book is on achieving distillation product purity at low cost without dwelling on complex mathematical descriptions.

Often, there is more than one way to achieve the desired results, so conflicts can develop about which is the best way to approach a subject and what creates the most value. The cost of acquiring resources and developing everyone's knowledge is generally too high. On the other hand, the risk of failing to meet production rates or product quality or of consuming excessive raw materials and energy has economic consequences for the business. Generally, the highest value is created by striking a balance.

## 1.3 Audience and Prerequisites

The material in this book can be useful for engineers, technicians, and plant operators concerned with the design, operation, and troubleshooting of process control systems for distillation columns. The course can also be useful for students who want to gain insights into the practical approach to distillation process control in industry and tuning control loops in a plant control room.

There are no specific prerequisites for taking this course. However, it would be helpful to have a basic understanding of the distillation unit operation and process control loop concepts. The only mathematics skills required are basic arithmetic and algebra.

## 1.4 Study Materials

This textbook is the only study material required. Additional references are provided in each chapter.

## 1.5 Organization and Sequence

This book is divided into 10 separate study chapters. Chapters 2, 3, and 4 deal with the distillation variables, and Chapter 5 covers distillation process control strategies. Chapter 6 describes some of the constraints on distillation variables and separation capabilities. Chapter 7 introduces the concepts that are critical to product quality and the measurements that evaluate performance criteria such as frequency of failure. Chapter 8 describes the concepts and nomenclature that are fundamental to PID control loops. Chapter 9 covers the concepts of tuning process controllers when they are operating in automatic output mode. Chapter 10 is about measuring the response of process variables when the controller is in manual output mode, that is, with no feedback from the process variable.

There are example problems and exercises for each chapter to test your understanding of the material. The solutions to all of the exercises are given in the appendix.

## 1.6 Course Objectives

When you have completed this entire book, you should be able to

1. Understand the manipulated variables and controlled variables for a distillation column
2. Understand the fundamentals of various strategies for controlling distillation columns
3. Understand many of the constraints and limits that cannot be exceeded in distillation control
4. Understand the measurements that are critical to product quality performance and the resultant frequency of failure
5. Understand how to tune a process controller when it is in automatic output mode and how to demonstrate the response of a process with the controller in manual output mode

## 1.7 Course Length

The organization of this book into chapters is designed for either classroom teaching or self-paced learning that can meet the needs and skill level of each student. Chapter 2 is helpful for understanding the nomenclature used throughout the book regarding the naming conventions for the distillation process streams. Chapter 7 can be studied alone for understanding the optimization of product quality performance and the frequency of failure. Chapters 9 and 10 can be studied separately for tuning control loops.

## References

1. Kister, H. Z., *Distillation Operation*, McGraw-Hill, New York, 1989.
2. Kister, H. Z., *Distillation Design*, McGraw-Hill, New York, 1992.
3. Buckley, P. S., Luyben, W. L., and Shunta, J. P., *Design of Distillation Column Control Systems*, ISA, Research Triangle Park, North Carolina, 1985.

*Chapter 2*

# Distillation Control Variables

This chapter provides you with a concept drawing of a distillation column with all of the streams labeled.

## Learning Objectives

When you have completed this chapter, you should be able to

1. Use the distillation column concept drawing for reference and for communication with others about the key variables and control elements
2. Recognize the names of the inlet and outlet streams used in a distillation system
3. Know the standard definition of reflux ratio

Distillation is the main unit operation in chemical engineering for the separation and purification of liquids and vapors. A feed mixture of chemicals can be separated into the more

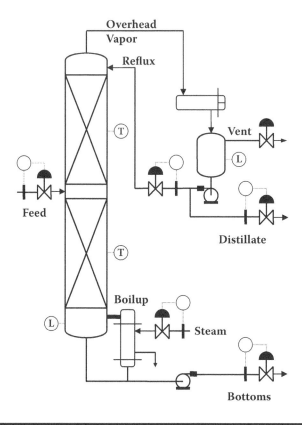

**Figure 2.1   Distillation column control.**

volatile components in a distillate stream at the top of a distillation column and the less volatile components in a bottoms stream (Figure 2.1). The inlet and outlet stream flow rates are the manipulated variables in distillation column control. A manipulated variable is usually the flow rate of a stream that has an automatic valve connected to a controller output signal.

## 2.1 Distillation Column Inlet Streams

For a continuous binary (two-component) distillation column, there are three streams entering the column:

1. Feed
2. Boilup
3. Reflux

## 2.2  Distillation System Outlet Streams

There are two main streams and one minor stream leaving a distillation system:

1. Distillate
2. Bottoms
3. Small vent stream of inert or light vapors leaving the reflux drum

## 2.3  Controlled Variables

A controlled process variable is usually a flow rate, pressure, or temperature signal from a sensor and transmitter that is used to provide the input to a controller. A controlled variable is controlled to a setpoint.

The main variables to be controlled in a distillation column are the following:

1. Distillation column pressure
2. Reflux drum liquid level
3. Column bottom liquid level
4. Separation power base
5. Material balance split

The pressure in a distillation column has an overriding influence on the control of the process. The pressure needs to be run as steady as possible because a reduction in pressure can cause a surge in vaporization. An increase in pressure can cause a surge in condensation.

The separation power base in the classic McCabe–Thiele[1] graphical model of a binary distillation column is established by the reflux ratio, $R/D$, which is the ratio of the reflux flow rate divided by the distillate flow rate. For example, with a distillation column that is fed 1,000 kg/h of feed that produces 85 kg/h of distillate with 425 kg/h of reflux, the reflux ratio is 425/85 = 5. A minimum reflux ratio is required to achieve the desired separation with an infinite number of theoretical stages. The maximum reflux ratio, called total reflux, with zero distillate flow rate can be used in design calculations to determine the minimum number of theoretical stages required to achieve a desired separation.

The desired material balance split, $D/F$, that is, the ratio of the distillate flow rate divided by the feed rate, is generally determined by the weight fraction of light components in the feed stream. For example, a distillation column for separating ethanol from water that produces 85 kg/h of distillate from 1,000 kg/h of feed would be running with a $D/F$ ratio of 85/1000 = 0.085.

## 2.4 Summary

This chapter presented a concept drawing of a distillation process that included flow rate sensors, liquid level sensors, temperature sensors, and process control valves. The process streams were identified and labeled. The standard definition of reflux ratio was presented, and the concepts of separation power and material balance split were introduced.

# Exercises

2.1 With a reflux ratio (*R/D*) of 4.0 and a material balance split (*D/F*) of 0.5, what is the ratio of reflux/feed?

2.2 If the reflux ratio (*R/D*) is 3.0, what is the ratio of reflux to overhead vapor (*R/O*)?

# Reference

1. McCabe, W. L., and Thiele, E. W., Graphical design of fractionating columns, *Ind. Eng Chem.*, 17, 605, 1925.

# Chapter 3

# Separation Power

The separation power in a distillation column is associated with the height of the column and the energy consumed to achieve the desired separation and purification of products.

## Learning Objectives

When you have completed this chapter, you should be able to

1. Understand the origin of separation power in distillation
2. Calculate the key impurity separation power
3. Conceptualize the control of separation power with the ratio of steam/feed or reflux/feed
4. Consider ways of minimizing energy consumption

## 3.1 Relative Volatility in One Theoretical Stage

The separation power in a distillation column comes from the thermodynamic vapor–liquid equilibrium relative volatility, the number of theoretical stages achieved by the column, and the energy consumed, that is, boilup and reflux flow rates, relative to the feed rate. The relative volatility in a single stage originates thermodynamically from the ratio of pure component vapor pressures and the ratio of activity coefficients in the liquid phase (Equation 3.1). The separation power is the ratio of component 1 to component 2 in the vapor phase divided by the ratio of component 1 to component 2 in the liquid phase. Nonidealities in the vapor phase are often insignificant below about 2 atm of pressure.

$$\alpha = \gamma_1 p_1 / \gamma_2 p_2 = (y_1/y_2)/(x_1/x_2) \qquad (3.1)$$

where:
$\alpha$ = relative volatility, alpha
$\gamma$ = activity coefficient in liquid phase, gamma
$p$ = pure component vapor pressure
$x$ = mole fraction in liquid phase
$y$ = mole fraction in vapor phase
subscript 1 = component 1
subscript 2 = component 2

For example, when 4.88 wt% ethanol in water is in equilibrium with 35.58 wt% ethanol in water vapor at atmospheric pressure, the relative volatility, $\alpha$, is equal to (35.58/64.42)/ (4.88/95.12) = 10.77.

When a binary system is ideal with no interaction, the activity coefficients in the liquid phase are 1.0 and the system obeys Raoult's law. One rough rule of thumb is that the ratio of pure component vapor pressures is equal to 1.036 raised to the power of the difference in boiling points in degrees Celsius. Komori and Ohe[1] reported a complete set of vapor–liquid equilibrium data for cyclohexane/*n*-heptane at atmospheric pressure as shown in Table 3.1. The atmospheric boiling point of *n*-heptane is 98.4°C, and the atmospheric boiling point of cyclohexane is 80.7, so the difference in boiling point is 17.7°C. The rough rule of thumb suggests that the ratio

**Table 3.1  Vapor–Liquid Equilibrium Data for Cyclohexane/*n*-Heptane at 760 mm Hg**

| T (°C) | Mole Fraction in Liquid | | Mole Fraction in Vapor | | Relative Volatility |
|---|---|---|---|---|---|
| | C6 | C7 | C6 | C7 | |
| 98.40 | 0.0000 | 1.0000 | 0.0000 | 1.0000 | |
| 97.72 | 0.0250 | 0.9750 | 0.0430 | 0.9570 | 1.752 |
| 97.17 | 0.0480 | 0.9520 | 0.0810 | 0.9190 | 1.748 |
| 96.51 | 0.0770 | 0.9230 | 0.1270 | 0.8730 | 1.744 |
| 95.51 | 0.1180 | 0.8820 | 0.1880 | 0.8120 | 1.731 |
| 93.91 | 0.1890 | 0.8110 | 0.2870 | 0.7130 | 1.727 |
| 93.41 | 0.2130 | 0.7870 | 0.3180 | 0.6820 | 1.723 |
| 92.47 | 0.2580 | 0.7420 | 0.3730 | 0.6270 | 1.711 |
| 91.26 | 0.3200 | 0.6800 | 0.4460 | 0.5540 | 1.711 |
| 90.45 | 0.3650 | 0.6350 | 0.4960 | 0.5040 | 1.712 |
| 89.79 | 0.4010 | 0.5990 | 0.5320 | 0.4680 | 1.698 |
| 88.10 | 0.4980 | 0.5020 | 0.6270 | 0.3730 | 1.694 |
| 86.75 | 0.5770 | 0.4230 | 0.6970 | 0.3030 | 1.686 |
| 85.17 | 0.6750 | 0.3250 | 0.7770 | 0.2230 | 1.678 |
| 84.06 | 0.7500 | 0.2500 | 0.8330 | 0.1670 | 1.663 |
| 83.49 | 0.7930 | 0.2070 | 0.8630 | 0.1370 | 1.644 |
| 82.81 | 0.8450 | 0.1550 | 0.9000 | 0.1000 | 1.651 |
| 81.97 | 0.9060 | 0.0940 | 0.9400 | 0.0600 | 1.625 |
| 80.70 | 1.0000 | 0.0000 | 1.0000 | 0.0000 | |

**Table 3.2  Mole Percent Cyclohexane in Liquid and Vapor at Total Reflux**

| Stage No. | Mol% C6 in Liquid | Mol% C6 in Vapor | Total Separation Power |
|:---:|:---:|:---:|:---:|
| 4 | 36.12 | 49.30 | 8.75 |
| 3 | 24.74 | 36.12 | 5.09 |
| 2 | 16.04 | 24.74 | 2.96 |
| 1 | 10.00 | 16.04 | 1.72 |

of pure component vapor pressures would be about $(1.036)^{17.7}$ = 1.87.

## 3.2 Separation Power with Multiple Stages

When a distillation column is run at total reflux for a long enough period of time to reach steady state, all of the vapor leaving a stage and going up the column will be coming back down the column as liquid from the stage above at the same concentration. For example, when the relative volatility is constant at 1.72, the separation power of four theoretical stages at total reflux is $(1.72)^4 = 8.75$ (Table 3.2).

The separation power in a distillation column can be defined by the ratio of light key to heavy key component in the distillate divided by the ratio of these components in the bottoms (Equation 3.2):

$$\text{Separation power} = (x_1/x_2)_D/(x_1/x_2)_B \qquad (3.2)$$

where:
    subscript D = distillate
    subscript B = bottoms

For example, a distillation column running at total reflux that achieves the equivalent of four theoretical stages of performance with cyclohexane and *n*-heptane would have 49.3 mol% cyclohexane in the distillate and 10.0 mol% cyclohexane in the bottoms, that is, the separation power = (49.3/50.7)/(10/90) = 8.75.

The Fenske[2] equation gives the minimum number of theoretical stages required to achieve a desired separation power at total reflux for a constant relative volatility (Equation 3.3):

$$N = \mathrm{Ln}\ [(x_1/x_2)_D/(x_1/x_2)_B]/\mathrm{Ln}\ \alpha \qquad (3.3)$$

where:

$N$ = Number of theoretical stages at total reflux

For example, Ln[(49.3/50.7)/(10/90)]/Ln 1.72 = 4.0 for the cyclohexane/*n*-heptane distillation at total reflux shown in Table 3.2.

In the control of distillation columns, the distillate composition is often close to 100% light component 1, and the bottoms is close to 100% heavy component 2, so the key impurity separation power can be calculated for control purposes from Equation 3.4.

$$\text{Key Impurity Separation Power} =$$

$$\frac{10,000}{(\text{wt\% heavy key in Distillate} \times \text{wt\% light key in Bottoms})} \qquad (3.4)$$

In other words, the key impurity separation power is indicated by the reduction of heavy key impurity from the distillate and the reduction of light key impurity from the bottoms. The maximum separation power that a distillation tower can achieve would be at total reflux, that is, total boilup and reflux with zero feed rate. One rule of thumb for an economical design of a distillation column is to use 2.0 times the minimum number of theoretical stages. This generally coincides with a design for about 1.3 to 1.5 times the minimum reflux ratio. A design of 4.0 times the minimum number of theoretical stages may use only 1.01 to 1.10 times the minimum reflux ratio.

## 3.3 Separation Power and Energy Consumption

For the design of a distillation column, there is an economic trade-off between the use of an extremely tall distillation column that runs with minimum energy consumption and a shorter column that requires higher energy consumption. The energy consumption is for heat to the reboiler and cooling for the condenser.

The separation of a feed mixture containing 50 wt% cyclohexane and 50 wt% *n*-heptane into a distillate of 99.9% cyclohexane and a bottoms of 0.1% cyclohexane requires a key impurity separation power of 1 million by Equation 3.4. With an infinite number of theoretical stages, a minimum reflux ratio of 2.40 is required to meet these product specifications. That is the lowest energy consumption possible for the desired separation by distillation.

The highest energy consumption is at total reflux, that is, when all of the boilup is returned as reflux, and there is no feed to the column and no distillate or bottoms. A computer simulation for this separation at total reflux required a minimum number of 27.8 theoretical stages. This was accomplished by setting the reflux/distillate ratio to 10 million in a computer simulation. This is the shortest column that can achieve the desired separation.

The desired separation can be achieved with 50 theoretical stages using a reflux ratio of 3.22, which is 1.34 times the minimum reflux ratio. This would meet upper specification limits (USLs) of 1,000 ppm *n*-heptane in the distillate and 1,000 ppm cyclohexane in the bottoms. However, the average key impurity concentrations in the products need to run with average impurity concentrations that are some distance below their product specification limits. This safety factor is needed, so the control system can shed disturbances without having the streams going above the specification limits of impurities.

The distance between the average key impurity concentration and the upper spec limit is the subject of product quality performance in Chapter 7. If the standard deviation ($\sigma$) is 10% of the upper specification limit, then the standard deviation for an upper specification limit of 1,000 ppm impurity is 100 ppm. A distillation column would need to produce an average key impurity concentration of 700 ppm impurity to keep the average DNS (distance from the nearest specification limit) at $3\sigma$.

The results of a steady-state simulation for a distillation column to separate a mixture of 50 wt% cyclohexane/$n$-heptane ($C_6$/$C_7$) into 99.93% $C_6$ in the distillate and 0.07% $C_6$ in the bottoms are shown in a McCabe–Thiele diagram (Figure 3.1). The simulation used 50 theoretical stages with a reflux ratio of 3.41 at atmospheric pressure with a feed temperature of 15°C.

Reducing the key impurity concentrations to 0.707 times the specification limit on each end of the distillation column

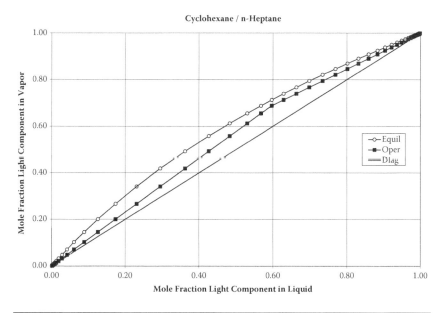

**Figure 3.1   McCabe–Thiele diagram.**

**Table 3.3   Energy Consumption to Double the Key Impurity Separation Power for C₆/C₇ Distillation with 50 Theoretical Stages**

|  | Minimum Separation Power | Design Separation Power |  |
|---|---|---|---|
| $X_D$ = | 0.9990 | 0.9993 | Weight fraction C6 in distillate |
| $X_B$ = | 0.0010 | 0.0007 | Weight fraction C6 in bottoms |
| $R/D$ = | 3.22 | 3.41 | kg reflux/kg distillate |
| $R/F$ = | 1.61 | 1.71 | kg reflux/kg feed |
| $Q_r/F$ = | 400 | 415 | btu reboiler duty/kg feed |

doubles the key impurity separation power, that is, from 1 million to 2 million in this case. This required the reboiler duty to be increased 4% and the reflux flow rate to be increased 6% per pound of feed (Table 3.3).

The separation power base for distillation control can be set either by the ratio of reflux/feed or by the ratio of boilup/feed. If the separation power base is set by the ratio of boilup/feed, then either the reflux or the distillate flow rate can be manipulated to control the distillate/feed material balance split. If the separation power base is set by the ratio of reflux/feed, then the steam, that is, boilup, flow rate can be manipulated to control the distillate/feed material balance split.

## 3.4 Summary

This chapter has derived the separation power for a distillation column from its origin in vapor–liquid equilibrium to the key impurity separation power for a distillation column with a stripping and a rectification section. The stripping section primarily removes light key impurity from the bottoms liquid stream, and the rectification section primarily removes heavy key impurity from the overhead vapor stream. The separation power for a distillation column is associated with the energy consumption, that is, steam and cooling water per pound of feed.

## Exercises

3.1 What is the key impurity separation power for a benzene and toluene binary distillation with 1% benzene in the bottoms and 98% benzene in the distillate?

3.2 What is the minimum number of theoretical stages required for a $C_3$ splitter to separate propylene from propane if the geometric mean relative volatility between the top and bottom is 1.12, and the desired separation is 99.5% propylene in the distillate and 1% propylene in the bottoms?

3.3 If a desired ethanol–water separation by distillation requires a minimum of 17 theoretical stages at total reflux, how many theoretical stages would be recommended for a first-pass design of a plant column using a computer simulation?

## References

1. Komori, T., and Ohe, S., Vapor-liquid equilibrium data for the system: Cyclohexane-*n*-heptane at atmospheric pressure, *Kagaku Kogaku Ronbunshu*, 16, 384, 1990.
2. Fenske, M. R., Fractionation of straight run Pennsylvania gasoline, *Ind. Eng. Chem.*, 24, 482, 1932.

# Chapter 4

# Distillate/Feed Material Balance Split

The split of the feed stream into a distillate stream and a bottoms stream can be referred to as the *material balance split*. Generally, the ratio of distillate/feed is about equal to the weight fraction of light components in the feed.

## Learning Objectives

When you have completed this chapter, you should be able to

1. Understand the importance of controlling the material balance split in distillation
2. Conceptualize the control of the material balance split with a temperature controller
3. Look for the MRT (most responsive temperature) point in the column with five different methods
4. Calculate the distillate/feed ratio from feed, distillate, and bottoms concentrations

## 4.1 Material Balances

A total material balance for a distillation column in Chapter 2, Figure 2.1, is shown in Equation 4.1. The individual component material balance is shown in Equation 4.2.

$$F = D + B \tag{4.1}$$

$$FX_F = DX_D + BX_B \tag{4.2}$$

where:
　　$F$ = feed stream (kg/h)
　　$D$ = distillate stream (kg/h)
　　$B$ = bottoms stream (kg/h)
　　$X_F$ = mass fraction of component in feed stream
　　$X_D$ = mass fraction of component in distillate stream
　　$X_B$ = mass fraction of component in bottoms stream

By combining Equations 4.1 and 4.2, the material balance split is shown to be directly related to the compositions:

$$\text{Material balance split} = D/F = (X_F - X_B)/(X_D - X_B) \tag{4.3}$$

The distillate/feed material balance split is determined by the concentration of light component in the feed, distillate, and bottoms. The weight fraction of feed flow that becomes distillate product is $D/F$, while the remaining fraction of the feed becomes bottoms product $1-D/F$. When the value of $X_F$ represents the weight fraction concentration of light components, the ratio of $D/F$ is nearly equal to $X_F$. This occurs because the weight fraction of "lights" in the bottoms, $X_B$, is near zero, and the weight fraction of lights in the distillate, $X_D$, is near one. When $D/F$ is greater than $X_F$, there will be "heavies" going over in the distillate. When $D/F$ is less than $X_F$, there will be lights going down in the bottoms.

For example, with 1,000 kg/h of feed, $F$, to an ethanol/water distillation column with 100 kg/h of distillate, $D$, that is rich in ethanol and 900 kg/h of bottoms, $B$, that is mostly water, the ratio of distillate to feed, $D/F$, is $100/1,000 = 0.10$.

If the feed stream concentration is 0.08 kg ethanol per kg of feed, that is, $X_F$, and the ethanol concentration in the bottoms, $X_B$, is 1,000 ppm, that is, 0.001 kg ethanol per kg of bottoms, and the distillate concentration, $X_D$, is 0.95 kg ethanol per kg of distillate, the ratio of distillate/feed, $D/F$, is equal to $(0.08 - 0.001)/(0.95 - 0.001) = 0.0832$ and $D = 0.083 \times 1,000 = 83$ kg/h. The bottoms flow rate is $1,000 - 83 = 917$ kg/h.

In graphical calculations, Equation 4.3 is known as the *inverse lever–arm rule*. A graph of the weight fraction of component 1 versus component 2 gives a straight line that connects the bottoms, feed, and distillate compositions. The feed flow rate is proportional to the length of the entire line called a *lever* $(X_D - X_B)$. The distillate flow rate is proportional to the length of the inverse lever arm $(X_F - X_B)$, and the bottoms flow rate is proportional to the length of the other inverse lever arm $(X_D - X_F)$. The $D/F$ ratio will be somewhere between 0.0 (no distillate and all bottoms) and 1.0 (all distillate and no bottoms).

One result of studying the material balance split can be to understand that the compositions and flow rates of both the distillate and bottoms streams interact with each other and cannot be independently controlled. If the feed flow rate and one of the product stream flow rates are fixed, then the flow rate of the other product stream must be the difference between them, or accumulation would occur. Compositions of the distillate and bottoms streams are directly linked by the material balances. Control of the distillate and bottoms quality can be achieved by controlling the separation power base and material balance split to achieve the desired compositions. The flow rates of the distillate and bottoms streams change as necessary to maintain the material balance and achieve the target compositions.

The feed to the distillation column in Figure 2.1 is split into a distillate stream and a bottoms stream. The fraction of feed going into distillate, that is, $D/F$, strikes a balance between distillate purity and bottoms purity. For example, with 50% lights in the feed, a $D/F$ ratio of 0.52 will force heavy impurities into the distillate. Similarly, a $D/F$ of 0.48 will force light impurities into the bottoms. A high $D/F$ ratio will result in all of the temperatures going up in a distillation column, but some temperature points will increase much more than others. For example, the top and bottom temperatures may not change much when the impurities are in the ppm range. However, the temperature point may respond quite significantly where the vapor concentration in the column is about 50% lights and 50% heavies. If the $D/F$ ratio is shifted up, then the vapor composition and temperature from the tray below will be shifted up into the tray of interest. Similarly, if $D/F$ is shifted down, the liquid composition and temperature from the tray above will be shifted down into the tray of interest.

## 4.2 Temperature Gradient per Theoretical Stage

A steady-state simulation for the distillation of 50 wt% cyclohexane in *n*-heptane with 50 theoretical stages at atmospheric pressure was run to study the temperature and vapor composition gradient above and below each theoretical stage (Figure 4.1). The feed is on Stage 23 from the bottom. The base case was run with 0.07% C7 (heavy) impurity in the distillate and 0.07% C6 (light) impurity in the bottoms. The difference between the temperature on the stage below and the stage above can be calculated for each stage and divided by 2. The temperature goes up almost linearly with the concentration of heavy key component in the vapor phase, so the concentration of light component was used from the stage above minus the stage below, that is, $(y_{n+1} - y_{n-1})/2$. The composition gradients for the top and bottom stage were

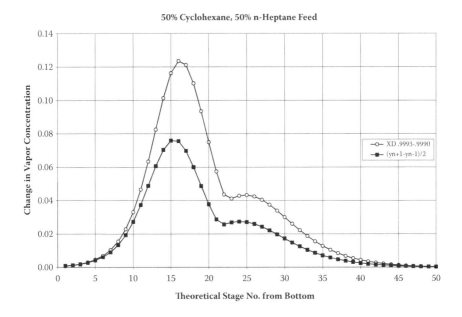

**Figure 4.1    Vapor concentration response from concentration gradient per theoretical stage and from increased *D/F*.**

calculated as the difference between the last two stages. This shows that the largest temperature gradient per theoretical stage is 7.6% of 17.7°C, that is, 1.3°C, at stage number 15 or 16 from the bottom.

For example, if at theoretical stage 15, temperature is 1.35°C hotter than Stage 16, and Stage 17 is 1.25°C cooler than Tray 16, the average response would be a shift of 1.30°C from the nearest tray.

## 4.3  Temperature Change from *D/F* Shift

The method reported by Tolliver and McCune[1] for finding the most responsive temperature point was also used for this cyclohexane/*n*-heptane system. A second computer simulation was run using the same reflux/feed ratio for the separation power base while shifting the *D/F* (distillate/feed) ratio up. The heavy key impurity concentration in the distillate was increased from 700 ppm to 1,000 ppm C7, that is, from 99.93 wt% to 99.90 wt% C6, while the light key impurity concentration in the bottoms went down from 700 ppm to 400 ppm C6. The results were similar to the tray temperature gradient method (Figure 4.1). The *D/F* shift gave a maximum temperature change of 12.4% of 17.7°C, or 2.2°C, at stage number 16 from the bottom as being the MRT (most responsive temperature), that is, the most responsive vapor composition point, in the column when there was a change in *D/F*. This small 300 ppm shift in material balance at the top and bottom of the column nearly shifted the stage 14 composition up into stage 16.

A shift in *D/F*, that is, distillate/feed ratio, material balance split has a greater effect on a small stream than on a large stream. For example, if the feed contains 10% lights, then the ratio of distillate/feed ratio would be about 0.10, so the concentration change and variability would be more significant in the small distillate stream. However, if the feed contains 90% lights, then the concentration change would be most significant in the small bottoms stream. Consequently, a study with two simulations should be run with a constant separation power base and a change of composition in the smaller of the two streams between the distillate and bottoms.

When the distillate heavy key impurity moved up from 0.0700% to 0.1000%, that is, a shift of 0.0300% C7, and the bottoms light key impurity moved down from 0.0700% to 0.0400%, that is, 0.0300% C6, the stage number 16 shifted up 12.4% C7. In this case, the stage 16 temperature moved 413 times more than the top temperature due to change in composition. Any significant temperature movement at the top or bottom of the column would be mostly due to changes in pressure.

The tray temperature gradient method requires only one steady-state simulation at design conditions, while the shift in material balance method requires two simulations with different product stream compositions. The maximum tray temperature gradient of 7.6% of 17.7°C or 1.35°C/tray occurs at stages 15 and 16 as shown in Figure 4.1. These observations show

that the temperature movement at a point in a distillation column is a result from the compositions above and below that
point in the column when the pressure is steady.

In addition to the two methods described for finding the
MRT (most responsive temperature) point in a distillation column, there are three additional methods that can be used without a computer simulation. First, if there are temperature points
in the column that are monitored, just look at the history record
to see which temperature point moves the most. Second, find
the temperature point in the column that is nearest to being
halfway between the top temperature and the bottom temperature. Third, look in the area of the column where the vapor
concentration would be about 50% light key component. For
example, if the feed contains 10% light key component, there is
a high probability that the MRT would be above the feed point.
Similarly, if the feed contains 80% light key, there is a high
probability that the MRT would be below the feed point.

Many of the disturbances to a distillation column come
from the feed stream changing in flow rate, temperature, or
composition. Other disturbances include changes in steam
header pressure, changes in cooling water temperature, and
changes in cooling water supply pressure. Also, changes in
ambient temperature such as day and night temperature cycles
and sudden rainstorms can cause disturbances. The MRT point
in the distillation tower generally responds first. Most of the
disturbances can be shed by process control action before
either of the end compositions go out of spec. For vacuum
towers, it is sometimes necessary to add pressure compensa-
tion to the MRT, so the control action is primarily for changes
in composition.

## 4.4 Summary

The multiple stages of counter-current contact with mass
transfer in a distillation column cause a temperature profile to

develop across the column. The highest temperature is found at the bottom of the column, while the lowest temperature is found at the top of the column. The temperature profile corresponds directly to the material balance split within the tower. There exists a point within the tower where the temperature will change the most as the material balance and the composition profile change. The temperature point that is most responsive to changes in material balance split tends to be near the point where the vapor composition is about 50% light key component and 50% heavy key component. The most responsive temperature point in the tower:

1. Responds the most to a change in material balance split
2. Can be anywhere in the tower
3. Tends to be near the mid-temperature (50% in vapor)
4. Can be identified by computer simulations
5. Can be seen in the control room by watching which temperature point moves the most
6. Can move within a range (operating window) while both distillate and bottoms compositions stay within spec.
7. Tends to be closer in the tower to the end with the smaller product stream

## Exercises

4.1  With a column distilling a binary mixture of cyclohexane and *n*-heptane near to total reflux with 10% cyclohexane in the bottoms and 49% cyclohexane in the distillate and 4.0 theoretical stages, which stage would have the most responsive temperature due to changes in *D/F* material balance split?

4.2  Calculate the *D/F* ratio if the feed concentration is 50% cyclohexane and 50% *n*-heptane, the distillate contains 0.1% *n*-heptane, and the bottoms contains 0.1% cyclohexane.

4.3 Is the MRT (most responsive temperature) point in the distillation column expected to be above or below the feed point when the liquid feed composition is 70% light key component?

4.4 Where is the MRT for a top-fed stripping column with 3% light key component in the feed?

## Reference

1. Tolliver, T. L., and McCune, L. C., Distillation control design based on steady-state simulation, *ISA Transaction*, Vol. 17, No. 3, 3–10, 1978.

## Chapter 5

# Distillation Control Strategies

The strategy of distillation control is open to many creative approaches because there are five or six main variables to manipulate and many possibilities for cascade, feed-forward ratios, and model-based computer control as well as conventional feedback control.

## Learning Objectives

When you have completed this chapter, you should be able to

1. Understand the four basic distillation control strategies
2. Consider more than one way to control a distillation column
3. Conceptualize the simultaneous control of separation power base with steam/feed ratio or reflux/feed ratio and the material balance split with a temperature controller
4. Look at the reflux and distillate flow rates and select the larger stream flow rate to manipulate for level control and the smaller stream for temperature control if steam/feed ratio is being used for the separation power base

**Table 5.1   Four Basic Control Strategies**

| Ratio That Establishes Separation Power Base | Manipulated Variable for D/F (Temperature) Control |
|:---:|:---:|
| Steam/feed | Distillate |
| Steam/feed | Reflux |
| Reflux/feed | Steam |
| Reflux/feed | Bottoms |

Tolliver and McCune[1] described four alternative material balance control schemes for distillation columns. Each scheme manipulates a different variable to control a temperature point in the column, that is, the *D/F* ratio, and this provides a structure for categorizing distillation control strategies (Table 5.1). There can be a number of variations of the four basic control schemes.

# 5.1  Column Pressure Control

One common strategy for controlling column pressure is to manipulate a valve in a vent line from the reflux drum. If necessary, another line and automatic valve can be added to inject inerts, for example, nitrogen, before the vent valve. If the distillate is taken as a vapor stream, then the condenser may need to be run as a partial condenser with temperature-controlled cooling liquid on the condenser.

# 5.2  Temperature Control with Distillate Flow Rate

In the manipulated distillate scheme, a column temperature controller manipulates a control valve in the distillate line. In other words, the column temperature is the process variable that is controlled to a setpoint, and the controller output is the signal to the control valve in the distillate line. The reflux drum level controller manipulates a valve in the reflux line, and the column base level controller manipulates a valve in the bottoms line. The feed rates to the column and reboiler steam are each on flow rate control. In one variation, there is a controller for the pressure drop across the trays that manipulates the valve in the reboiler steam line. The trays are used as the vapor flow rate sensing device. However, it is preferred to measure the steam flow rate to the reboiler and simply monitor the tower pressure drop, especially for packed towers. With this scheme, the separation power base is derived from the ratio of steam/feed. The distillate/feed material balance split is maintained by the MRT (most responsive temperature) point controller.

In some cases, the temperature controller for this manipulated distillate scheme has not been tuned well, so a plant operator changes the controller to manual output. Subsequently, the operators may leave the controller in manual output and simply adjust the distillate valve position manually to keep the MRT in a desired range or to keep the analysis of grab samples of distillate and bottoms in certain ranges. In other cases, the control system is actually designed to run with a manual setpoint for a distillate flow rate controller. The plant operators are required to grab samples once or twice during an 8-hour shift and get them analyzed immediately. There may not be any temperature measurements available in the column for automatic control of the $D/F$ material balance split. This requires significant operator intervention to shed disturbances that come at the tower, so one operator can only run a few towers at once. This type of manual control is sometimes required if the entire column operates with a low concentration range of impurity. In one case, the feed contained only 200 ppb impurity that had to be removed down to 5 ppb at one end of the column and concentrated to 10,000 ppb (10 ppm) concentration at the other end.

The manipulated distillate scheme is easy to start up on total reflux simply by closing the distillate flow valve. The level controller for the reflux drum manipulates the reflux flow rate to the column. This scheme of controlling temperature by manipulating the distillate flow rate works well for systems that run with high reflux ratios, especially when the reflux/distillate is greater than 5. This is compatible with a general rule of thumb that the larger stream should be used for level control and the smaller stream for column temperature control. This scheme also works well for towers operating at a boilup rate near to flooding because the vapor traffic can run quite steady. Environmental disturbances such as day and night temperature changes, or a rain storm, are shed reasonably well because the primary effect is on internal reflux in the column.

Another variation on the manipulated distillate scheme is to add a cascade slave control loop for the distillate flow rate as shown in Figure 2.1. The temperature control loop then manipulates the setpoint for the slave distillate flow control loop. Similarly, a slave flow control loop can be used for the reflux flow rate and another for the bottoms flow rate. At one point in time, the use of cascade control loops was called *advanced control*; it was compared to SISO (single input single output) control, because it required the addition of more hardware PID controllers. Modern computer control systems simply require the addition of software code to the program when the flow sensors are present.

Another variation on the manipulated distillate scheme is to use a setpoint for the steam/feed ratio to establish the separation power base. Generally, the feed flow rate signal should be lagged with an 8 to 20 min capacitance lag (filter), so the steam flow is proportional to a trailing average of the feed rate. Sometimes, this falls into the category of model-based predictive control because the McCabe–Thiele model, or a computer simulation model shows that the separation power base can be established by the steam/feed ratio as shown in Table 3.3.

A less common variation of the distillate scheme is used on the occasion where the first column in a distillation train is for removing light components, and the bottoms flow rate needs to be very steady as the flow rate to a large diameter fractionator as the second tower. In this case, the bottoms flow rate is set manually for the controller that manipulates the bottoms valve. The column-base level controller in the first-column manipulates the feed rate to the first column.

## 5.3 Temperature Control with Reflux Flow Rate

In the reflux scheme, a column temperature controller manipulates a control valve in the reflux line. The reflux drum level controller manipulates a valve in the distillate line. The column base level controller manipulates a valve in the bottoms line. The feed and reboiler steam are each on flow rate control. In some cases, there is a controller for the pressure drop across the trays that manipulates the valve in the reboiler steam line. However, it is preferred to use a steam flow rate controller and simply monitor the tower pressure drop. With this scheme the separation power base is derived from the ratio of steam/feed. The distillate/feed material balance split is maintained by the MRT point controller.

The reflux scheme can be difficult to start up because initially there may not be enough light components accumulated in the reflux drum, and the column temperature may be too hot; so, the controller may want more reflux flow than is available. This can pump the reflux drum level down until the distillate flow stops and then proceed to pump the reflux drum empty. However, this situation can be handled with computer control by using a low-level constraint control that will constrain the reflux flow rate to maintain a low-level constraint setpoint until the column temperature is low enough, so the temperature controller calls for less reflux. This reflux scheme is recommended when the reflux/distillate ratio is less than

0.8. The distillation column can be run very close to flooding because the boilup and vapor traffic can be run steady at maximum rate. This manipulated reflux scheme can also handle environmental disturbances well such as day and night temperature variations and rain storms.

Another variation on the manipulated reflux scheme is to add a cascade slave control loop for the reflux flow rate as shown in Figure 2.1. The temperature control loop then manipulates the setpoint for the slave reflux flow control loop. Similarly, a slave flow control loop can be used for the distillate flow rate and another for the bottoms flow rate.

Another variation on the manipulated reflux scheme is to use a setpoint for the steam/feed ratio to establish the separation power base. Generally, the feed flow rate signal should be lagged with an 8 to 20 min capacitance lag (filter), so the steam flow is proportional to a trailing average of the feed rate. Sometimes, this falls into the category of model-based predictive control because the McCabe–Thiele model or a computer simulation model shows that the separation power base can be established by the steam/feed ratio as shown in Table 3.3.

A less common variation on this control scheme is used on the occasion where the first column in a distillation train is for removing light components, and the bottoms flow rate needs to be very steady as the flow rate to a large diameter fractionator as the second column. In this case the bottoms flow rate from the first column is the manually adjusted setpoint for the controller that manipulates the bottoms valve. The columnbase level controller for the first column manipulates the feed rate to the first column.

## 5.4 Temperature Control with Boilup (Steam Flow Rate)

In the boilup scheme, a column temperature controller manipulates a control valve in the steam line to the reboiler. The

reflux is on flow rate control. The reflux drum level controller manipulates a valve in the distillate line. The column-base
level controller manipulates a valve in the bottoms line. With
this scheme, the separation power base is derived from the
ratio of reflux/feed. The distillate/feed material balance split is
maintained by the MRT point controller.

The origins of the boilup scheme may go back to the days
when the feed flow rate and reflux flow rate to a distillation
column were controlled by manual valves and glass rotame-
ters in the control room. A pneumatic temperature controller manipulated an automatic valve in the steam line to the
reboiler. This may be difficult to imagine by those who have
only experienced modern-day equipment.

The manipulated boilup scheme is still used in many cases
and especially for top-fed stripping columns that have no
reflux stream and no rectification section above the feed. A
column temperature controller manipulates the steam flow rate
to the reboiler. The MRT point for a stripper is often the overhead vapor temperature leaving the column. If the overhead
vapor temperature is close to the bottoms temperature, then
an overhead vapor flow rate controller can be used to manipulate the steam flow rate.

The manipulated boilup scheme is quite easy to start up
after some liquid is accumulated in the reflux drum. However,
this scheme does not shed environmental disturbances very
well. When a sudden rain storm hits the distillation tower, the
temperature disturbance is shed by the temperature controller increasing the steam flow rate to the reboiler. This can be
a problem if the distillation tower needs to be running at its
maximum capacity near to flooding.

Another variation on the manipulated boilup scheme is to
use a setpoint for the reflux/feed ratio to establish the separation power base. Generally, the feed flow rate signal should
be lagged with an 8 to 20 min capacitance lag (filter), so the
reflux flow is proportional to a trailing average of the feed
rate. Sometimes, this falls into the category of model-based

predictive control because the McCabe–Thiele model or a computer simulation model shows that the separation power base can be established by the reflux/feed ratio; see, for example, Chapter 3, Table 3.3.

## 5.5 Temperature Control with Bottoms Flow Rate

In the bottoms scheme, a column temperature controller manipulates a control valve in the bottoms line. The reflux drum level controller manipulates a valve in the distillate line. The feed and reflux are on flow rate control. The column base level controller manipulates the valve in the steam line to the reboiler. With this scheme, the separation power base is derived from the ratio of reflux/feed. The distillate/feed material balance split is maintained by the MRT point controller.

The manipulated bottoms flow rate scheme is not used very frequently because of problems with the column base level control using steam. When a thermosiphon reboiler is used and the steam flow is increased, there is usually a reversal in the column base level response. The column level first rises and then falls. The usual case for using this scheme is when the feed concentration is 95% light key or more. In other words, most of the feed is distilled overhead from a few percent heavies or tars. The MRT point is usually in the reboiler. A variation on the scheme is to put the steam on flow rate control and let the column base level controller manipulate the feed rate to the column, similar to the way a maple syrup evaporator is run.

A variation on the bottoms scheme is to add a cascade slave control loop for the bottoms flow as shown in Figure 2.1. The temperature control loop then manipulates the setpoint for the slave bottoms flow control loop. Similarly, a slave flow control loop can be used for the distillate flow rate and another for the steam flow rate.

## 5.6 Side Draw Flow Rate

There are a number of cases where a side draw from a distillation column is used. If a liquid side draw flow rate is a small stream primarily for purging a concentration bubble of an impurity, the control scheme may be simply a matter of setting a small fixed purge rate. However, if a pasteurizing section is used at the top of a column, the distillate stream may be a very small lights purge stream with the liquid returning to the column being close to total reflux. The small distillate purge can be set at a fixed flow rate. The liquid side draw may be the main product, so the side draw flow rate may be manipulated by a temperature controller much like the manipulated distillate scheme described earlier.

When a side draw is used on a distillation column, there is another degree of freedom introduced into the control scheme. There are actually two material balance splits to keep in balance. One is the ratio of distillate/side draw, and the other is the ratio of side draw/bottoms. The separation power base can be set by the ratio of steam/feed, and then the distillate flow rate can be manipulated by a temperature controller for the MRT point above the side draw. The side draw flow rate can be manipulated by the second temperature controller for the MRT point below the side draw.

The side draw can be very pure when a DWC (Dividing Wall Column) is used (Figure 5.1). In that case, there is another design variable, which is the ratio of liquid to the feed side of the wall divided by the liquid to the product side of the wall.

A vapor side draw from a distillation column is not used very frequently, but, when it is, a large valve in the vapor side draw line can be used to control the side draw flow rate. However, the control is very sensitive when such a large valve is used. An alternative can be to use a fixed large valve position and a manipulated small valve position in parallel for control.

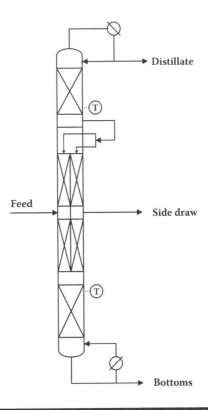

**Figure 5.1   Dividing wall column.**

## 5.7 Distillate Vapor Flow Rate

There are cases where a vapor stream from the top of a distillation column is needed as the distillate. In that case, the distillate vapor flow rate can be manipulated to control the column pressure and the condenser can be controlled as a partial condenser to produce the reflux. The heat duty for a condenser is a function of the heat transfer surface area and the temperature difference between the cooling medium and the condensing vapors. One approach is to recirculate the cooling fluid and control the temperature of the coolant for a variable temperature difference between the condensing vapors and the coolant. Another approach can be to use a condenser with the bottom of the tubes flooded, and the liquid condensate rate can be manipulated to allow a variable condenser area to be exposed to the vapors. When the partial condenser is cooled by an evaporating refrigerant, the liquid refrigerant level can be manipulated for control of the condenser heat duty.

## 5.8  Summary

In this chapter, four basic distillation control schemes were introduced along with a number of variations on each scheme. The concept was introduced that there can be more than one way to successfully control a distillation column. In some schemes, the separation power base is controlled by the ratio of steam/feed and then the distillate flow rate or reflux flow rate can be manipulated to control an MRT point in the distillation column. In other schemes, the separation power base is controlled by the ratio of reflux/feed, and the steam to the reboiler or the bottoms flow rate can be manipulated to control an MRT point in the column. Control of reflux drum level and column base level was presented as basic to all control schemes. Control of column pressure was considered to have an overriding effect on the stability of distillation column control.

## Exercises

5.1  What control strategy would be recommended for a top-fed stripping column?

5.2  What control strategy would be recommended for a batch rectification column that starts up at total reflux?

5.3  What control strategy would be recommended for a reflux ratio of 0.1?

# Reference

1. Tolliver, T. L., and McCune, L. C., Distillation control design based on steady-state simulation, *ISA Transaction*, Vol. 17, No. 3, 3–10, 1978.

# *Chapter 6*

# Constraints

For the operation and control of a distillation column, there are a number of constraints that must not be violated. Some of the constraints are mechanical and hydraulic limits, and others are mass transfer separation capability limits.

## Learning Objectives

When you have completed this chapter, you should be able to

1. Understand the maximum constraints for pressure drop and flooding, boilup, reflux, feed, and distillate flow rates
2. Understand the mass transfer rate limitations in the stripping section and the rectification section of a distillation column

## 6.1 Mechanical and Hydraulic Constraints

The most familiar mechanical constraint in the distillation system is the pressure drop and hydraulics associated with flooding. The main process variable that affects flooding is the vapor traffic in the column. This is driven by the boilup rate from the reboiler and creates a pressure drop across the trays or packing in the column. Above a certain pressure drop, liquid is held up in the packing or in the downcomers or trays. This condition is also associated with a maximum liquid flow that can be allowed for a column. Also, above certain velocities, the vapor can entrain liquid droplets and reduce tray efficiency or packing efficiency because of backmixing. This can limit or constrain the maximum separation power that is achievable.

All of the process flow rate variables are constrained because of the trim size in the control valves and may also be constrained because of the range of the flow rate sensors. Also, the liquid level in the reflux drum and the column base must be constrained to run between 0% and 100% of the level signals.

Another unique constraint can occur when the system to be separated has a high relative volatility, for example, 10 to 50, and the column is designed with, say, four times the minimum number of theoretical stages required at total reflux. The distillation column may be designed this way because one of the components is highly toxic and the project design team wants plenty of safety margin for removing the toxic component. Or, the condenser may need to run extremely cold or with vapor compression, so the project team wants to be certain that an absolute minimum of condenser duty is used. In these cases, the column may be able to run in the range of 1.01 times the minimum reflux ratio. However, the problem is that many of the extra stages in the column operate at a pinch condition around the feed tray. In other words, there is a series of trays that run at the same concentration and temperature. The inventory of material on these trays in the column can isolate one end of the column from the other end for a long period of time regarding process control of the $D/F$ material balance split. In this rare case, the solution could be to install and run two temperature control loops. The boilup is manipulated by a controller for a temperature point in the stripping section, and the distillate or reflux is manipulated by another controller for a temperature point in the rectification section.

## 6.2 Heat Transfer Constraints

The amount of boilup vapor produced from the reboiler is constrained by the heat transfer surface in the reboiler and the fouling in combination with the temperature of the heating medium compared to the boiling temperature. The amount of condensation is constrained by the amount of heat transfer surface in the condenser and fouling in combination with the temperature of the cooling medium compared to the bubble point of the overhead vapor stream at the column pressure.

## 6.3 Stripping Mass Transfer Constraint

There is also another constraint, or at least a point of diminishing return, based on the limitation of the rate of mass transfer that can take place in a distillation column as shown by Fair[1] and Kohl.[2] The stripping of volatile components in the stripping section below the feed point in a column is driven by the stripping factor of 1.0 or greater for the light key impurity, Equation 6.1. A stripping factor below 1.0 will be equilibrium pinched in the stripping section. A stripping factor above about 2.0 will be mass transfer limited for stripping solute from the liquid phase.

$$S = Vy/Lx \qquad (6.1)$$

where:
   $S$ = stripping factor, or lambda $(\lambda)$
   $V$ = vapor flow rate

$L$ = liquid flow rate

$x$ = concentration of solute in liquid phase

$y$ = concentration of solute in vapor phase at equilibrium

When dilute solute concentrations are stripped with a pure vapor, the number of theoretical stages can be calculated by Equation 6.2.

$$\text{NTS} = \text{Ln}[(x_F/x_B)(1 - 1/S) + 1/S]/\text{Ln } S \qquad (6.2)$$

where:

NTS = number of theoretical stages

$x_F$ = concentration of light key solute in liquid leaving feed stage

$x_B$ = concentration of light key solute in bottoms liquid

This model suggests that an infinite stripping factor can achieve any desired separation, that is, $x_F/x_B$, with zero theoretical stages. Clearly, that is not a good model of mass transfer at high stripping factors. The mass transfer in a stripping section becomes mass transfer rate limited in the liquid phase. In other words, even with an infinite stripping factor with a pure stripping gas, there will still be a required amount of contact time for desorption of the light impurity solute out of the liquid phase from $x_F$ down to $x_B$. The liquid phase mass transfer unit based on overall driving force in liquid phase concentrations, $N_{OL}$, gives a more realistic model of mass transfer at high stripping factors (Equation 6.3):

$$N_{OL} = \{\text{Ln}[(x_F/x_B)(1 - 1/S) + 1/S]\}/\{1 - 1/S\} \qquad (6.3)$$

where:

$N_{OL}$ = number of mass transfer units based on overall driving force in liquid phase concentrations

When the stripping factor is infinite, the number of mass transfer units can be calculated by Equation 6.4:

$$N_{OL} = Ln(x_F/x_B) \tag{6.4}$$

When the stripping factor is equal to 1.0, the values of NTS and $N_{OL}$ are identical (Equation 6.5).
For $S = 1.0$:

$$N_{OL} = NTS = x_F/x_B - 1 \tag{6.5}$$

In other words, reducing a light solute in the liquid from 1,000 ppm to 1 ppm requires 999 theoretical stages or mass transfer units when the stripping factor is 1.0. Table 6.1 shows a comparison of the number of theoretical stages and the number of liquid transfer units required for reducing the solute

**Table 6.1   Number of Theoretical Stages and Liquid Mass Transfer Units for Stripping 1000 ppm Light Impurity down to 1 ppm**

| $S$ | NTS | $N_{OL}$ |
|---|---|---|
| 1.0 | 999.00 | 999.00 |
| 1.1 | 47.42 | 49.72 |
| 1.2 | 28.09 | 30.73 |
| 1.3 | 20.75 | 23.59 |
| 1.5 | 14.33 | 17.43 |
| 2.0 | 8.97 | 12.43 |
| 5.0 | 4.15 | 8.36 |
| 10.0 | 2.95 | 7.56 |
| 20.0 | 2.29 | 7.22 |
| 100.0 | 1.50 | 6.97 |
| $\infty$ | 0.00 | 6.91 |

concentration by a factor of 1,000 with a pure stripping gas. The number of theoretical stages and transfer units are nearly the same for stripping factor values from 1.0 to 1.5. But the mass transfer becomes severely rate limited at higher stripping factors. If the stripping factor is 5 or higher in a distillation column because the relative volatility is high, then increasing the boilup will not increase the separation power by any significant amount in the stripping section.

When a computer simulation is used for a distillation column, the number of liquid transfer units can be calculated for each theoretical stage in the stripping section with Equation 6.6 using the stripping factor for the light key impurity in each stage, where NTS = 1:

$$N_{OL} = NTS\ (Ln\ S)/(1 - 1/S) \qquad (6.6)$$

## 6.4 Rectification (Absorption) Mass Transfer Constraint

A mass transfer constraint can also occur in the rectification (absorption) section of a distillation column when the value of $S$ falls below about 0.5 for the heavy key impurity in the vapor phase. When dilute, heavy, key solute concentrations are absorbed with a pure liquid, the number of theoretical stages can be calculated by Equation 6.7:

$$NTS = Ln[(y_F/y_D)(1 - S) + S]/Ln\ (1/S) \qquad (6.7)$$

where:
> NTS = number of theoretical stages
> > $y_F$ = concentration of heavy key solute in vapor leaving feed tray
> > $y_D$ = concentration of heavy key solute in distillate (overhead vapor)

This model suggests that a stripping factor of zero can achieve any desired separation, that is, $y_F/y_D$, with zero theoretical stages. Clearly, that is not a good model of absorption (enriching) mass transfer at low stripping factors. The mass transfer in a rectification (absorption) section becomes mass transfer rate limited in the vapor phase. In other words, even with a stripping factor of zero and a zero concentration of heavy key solute in the liquid entering the absorption section, there will still be a required amount of contact time for absorption of the heavy key impurity solute out of the vapor phase from $y_F$ down to $y_D$. The gas phase mass transfer unit based on overall driving force in gas phase concentrations, $N_{OG}$, gives a more realistic model of mass transfer at low stripping factors (Equation 6.8):

$$N_{OG} = \{Ln[(y_F/y_D)(1 - S) + S]\}/\{1 - S\} \qquad (6.8)$$

where:

$N_{OG}$ = number of mass transfer units based on overall driving force in gas phase concentrations

When the stripping factor is zero, the number of mass transfer units can be calculated by Equation 6.9:

$$N_{OG} = Ln(y_F/y_D) \qquad (6.9)$$

When the stripping factor is equal to 1.0, the values of NTS and $N_{OG}$ are identical (Equation 6.10):
   For $S = 1.0$:

$$N_{OG} = NTS = y_F/y_D - 1 \qquad (6.10)$$

In other words, reducing a heavy key solute in the vapor from 1,000 ppm to 1 ppm requires 999 theoretical stages or mass transfer units when the stripping factor is 1.0. Table 6.2 shows

**Table 6.2 Number of Theoretical Stages and Gas Mass Transfer Units for Rectification (Absorption) of 1,000 ppm Heavy Impurity down to 1 ppm**

| $S$ | $NTS$ | $N_{OG}$ |
|------|--------|--------|
| 1.00 | 999.00 | 999.00 |
| 0.91 | 47.42 | 49.72 |
| 0.83 | 28.09 | 30.73 |
| 0.77 | 20.75 | 23.59 |
| 0.67 | 14.33 | 17.43 |
| 0.50 | 8.97 | 12.43 |
| 0.20 | 4.15 | 8.36 |
| 0.10 | 2.95 | 7.56 |
| 0.05 | 2.29 | 7.22 |
| 0.01 | 1.50 | 6.97 |
| 0.00 | 0.00 | 6.91 |

a comparison of the number of theoretical stages and the number of gas transfer units required for reducing the heavy key solute concentration by a factor of 1,000 with a pure absorption liquid. The number of theoretical stages and transfer units are nearly the same for stripping factor values from 1.0 to 0.5. But the mass transfer becomes severely rate limited in the gas phase at lower stripping factors for the heavy key in the rectification section. If the stripping factor is 0.2 or lower in the rectification section because the relative volatility is high, then increasing the reflux will not increase the separation power by any significant amount.

When a computer simulation is used for a distillation column, the number of gas transfer units can be calculated for each theoretical stage in the rectification section with Equation

6.11 using the stripping factor for the heavy key impurity in each stage, where NTS = 1:

$$N_{OG} = \text{NTS} \, (\text{Ln } S)/(S - 1) \qquad (6.11)$$

Distillation of the cyclohexane–*n*-heptane system does not get into constrained mass transfer rate limited conditions, because the relative volatility is not high enough. The stripping factor for cyclohexane does not go above 1.75 in the stripping section, and the stripping factor of *n*-heptane does not go below 0.6 in the rectification section even at total reflux. In that case, the use of theoretical stages for design is reasonable.

## 6.5  Summary

In this chapter, all of the process flow rates were considered to be constrained by zero flow and the maximum flow allowable by the valve size and span of the flow measurement. The main column constraint because of flooding is associated with the vapor traffic and pressure drop across the trays or packing in the distillation column. The mass transfer rate limit for stripping light key impurity from the bottoms stream was presented. The mass transfer rate limit for absorbing heavy key impurity from the overhead vapor stream was also presented.

## Exercises

6.1  If the reflux flow to a distillation column uses cascade control, the span of the flow meter is 20,000 kg/h, and the reflux valve will allow 25,000 kg/h of flow when it is wide open, what is the reflux flow rate at the constraint limit?

6.2 With process water fed to the top of a stripping column at its boiling point and stripped with a ratio of 0.02 kg steam/kg water to remove 1,000 ppb of perchloroethylene down to 2 ppb, how many theoretical stages will be required if the relative volatility of perchloroethylene to water is 39,000?

6.3 With process water fed to the top of a stripping column at its boiling point and stripped with a ratio of 0.02 kg steam/kg water to remove, 1,000 ppb of perchloroethylene down to 2 ppb, how many liquid mass transfer units will be required if the relative volatility of perchloroethylene to water is 39,000?

6.4 A vent gas stream of air contains 200 ppm hydrochloric acid and is to be fed to an absorber to be cleaned to 0.1 ppm acid. A high molar flow rate of water is used with caustic that is 10 times the amount of acid present so the equilibrium concentration of HCL in the vapor is zero. How many gas mass transfer units will be required?

6.5 A computer simulation for the distillation of ethanol and water shows that the second theoretical stage from the bottom would operate with a stripping factor of 3.0. How many liquid phase mass transfer units are equivalent to that theoretical stage?

6.6 The water vapor from an evaporator contains 0.5% diethylene glycol, and this vapor is fed to an absorption column that runs with a reflux ratio of 0.1. How many gas phase mass transfer units will be required to reduce the concentration of glycol to 1 ppm in the water if the equilibrium concentration of diethylene glycol in liquid water is 350 times higher than the vapor phase?

# References

1. Fair, J. R., Distillation, Chapter 5 in *Handbook of Separation Process Technology*, Rousseau, R. W., ed., John Wiley & Sons, New York, 1987.
2. Kohl, A. L., Absorption and stripping, Chapter 6 in *Handbook of Separation Process Technology*, Rousseau, R. W., ed., John Wiley & Sons, New York, 1987.

# Chapter 7

# Optimizing Product Quality Performance

The goal is to continually provide product quality that meets customer needs and expectations at a price they are willing to pay and a manufacturing cost/unit that generates maximum economic return.

## Learning Objectives

When you have completed this chapter, you should be able to

1. Understand the measurements that are critical to product quality
2. Quantify the quality performance with statistical measurements, and understand the frequency of failure
3. Optimize product quality performance to generate maximum economic return

## 7.1 Quality Performance Measurement

One measure of the quality of distillate and bottoms from a distillation column is the concentration of impurities present. The rectification section of a distillation column above the feed point is primarily for removing heavy impurities from the distillate. So, the quality of the distillate stream can be measured by the concentration of heavy key impurity. The distillate quality specification can be an upper specification limit (USL) of heavy key impurity in the stream. If there are lighter components present, they will essentially all appear in the distillate.

Similarly, the stripping section of the distillation column below the feed point is primarily for removing light impurities from the bottoms stream. The quality of the bottoms stream can be measured by the concentration of light key impurity. The word *impurity* is simply used to indicate low concentrations. The actual light key impurity component left in the bottoms may be the product being sold and may represent a yield loss. The bottoms quality specification can be an upper specification limit (USL) for the light key impurity in the stream. Essentially, all heavier components will appear in the bottoms. A failure can be indicated by the impurity concentration exceeding the upper specification limit.

The concentration of key impurities in samples of distillate or bottoms can be studied with conventional statistical calculations, for example, average concentration and standard deviation. There can be common cause variability from everyday disturbances that occur to a distillation column and assignable cause variability from known disturbances such as startup, loss of steam, etc. The main emphasis of statistical measurement of product quality is on common cause variability.

## 7.2  Frequency of Failure

The quality performance of a stream from a distillation column can be evaluated by the average impurity concentration in the stream and the distance from the nearest specification limit (DNS) compared to the standard deviation (Equation 7.1). The value of DNS is three times the process capability index, $C_{pk}$. The DNS value is easy to communicate in the number of sigma units:

$$DNS = (USL - X_a)/\sigma \qquad (7.1)$$

where:
    DNS = distance from nearest specification limit
    USL = upper specification limit
    $X_a$ = average key impurity concentration
    σ = standard deviation (sigma)

When the concentration of impurity follows a normal distribution in samples, the frequency of failure rate can be calculated (Table 7.1). These values are taken from a single side of the normal Gaussian error distribution, which can be found in statistics books such as the one by Montgomery and Runger.[1]

When samples of distillate and bottoms are taken once every 8-hour shift, there are 21 samples per week and 1,095 samples per year. If the samples follow a normal distribution and the average impurity concentration is 2σ, that is, DNS = two standard deviations, below the upper specification limit (USL) there would be one shift sample out of spec about every 2 weeks. If the average were 3σ below the upper specification limit, there would be one shift sample out of spec about every 9 months from common cause variability. If the average were 4σ below the upper specification limit, there would be one shift sample out of spec about every 27 years.

**Table 7.1   DNS versus Frequency of Failure Rate for a Normal Distribution**

| DNS ($\sigma$) | Fraction of Samples Failing to Meet Specs |
|---|---|
| −0.5 | 7/10 |
| 0.0 | 1/2 |
| 0.5 | 1/3 |
| 1.0 | 1/6 |
| 1.5 | 1/15 |
| 2.0 | 1/44 |
| 2.5 | 1/161 |
| 3.0 | 1/741 |
| 3.5 | 1/4,290 |
| 4.0 | 1/31,400 |
| 4.5 | 1/290,000 |

Figure 7.1 shows a normal distribution of sample analyses with 1,000 ppm USL for heavy key impurity in the distillate and for light key impurity in the bottoms and 100 ppm standard deviation ($\sigma$). The DNS (distance from the nearest specification) is $3\sigma$ in each case.

## 7.3  Optimize MRT between Distillate and Bottoms Quality

Figure 7.2 shows the result of a high $D/F$ and high MRT (most responsive temperature) where the distillate quality performance DNS is reduced to $2\sigma$ and the bottoms quality performance DNS is increased to $4\sigma$. The frequency of failure rate has increased in the distillate stream because the column

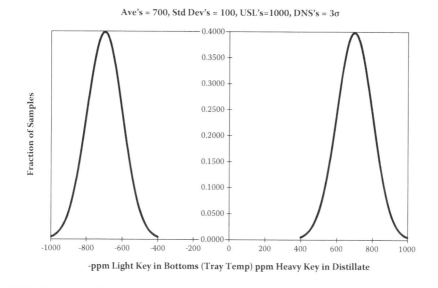

**Figure 7.1    Quality performance in distillation column.**

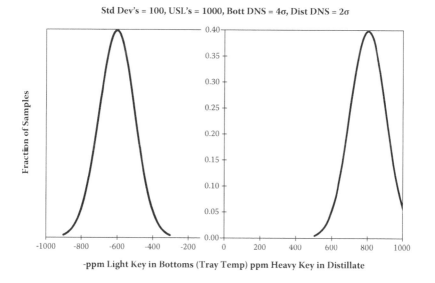

**Figure 7.2    High tray temperature in distillation column.**

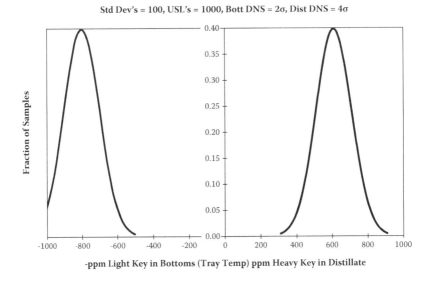

Std Dev's = 100, USL's = 1000, Bott DNS = 2σ, Dist DNS = 4σ

**Figure 7.3   Low tray temperature in distillation column.**

temperature is high and more heavy key impurity is in the distillate while the bottoms stream is being overpurified.

A similar but opposite response can also be visualized when the *D/F* and MRT are low, which leaves more light key impurity in the bottoms and overpurifies the distillate as shown in Figure 7.3.

## 7.4  Avoid Excessive Use of Steam

One way of reducing the failure rate is to increase the energy consumption; this will increase the DNS, that is, increase the distance between the upper specification limits and the average key impurity concentrations as shown in Figure 7.4. This is, of course, contingent on none of the variables or mass transfer being at a constraint limit. However, the use of excessive steam and cooling just to overpurify the products reduces the economic return from a distillation column and reduces the capacity because of the higher vapor traffic used per pound of product.

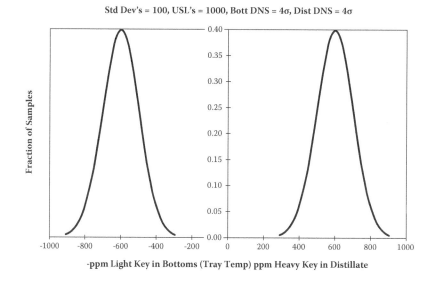

Std Dev's = 100, USL's = 1000, Bott DNS = 4σ, Dist DNS = 4σ

Figure 7.4    **High steam consumption in distillation column.**

# 7.5  Reduce Variability

One of the best ways of improving product quality at both ends of the distillation column is to reduce variability, as shown in Figure 7.5; that will reduce the failure rate. Robinson[2] described the benefits from improved dryer control, such as reduced variability in product moisture content and reduced energy consumption because the average concentration of moisture in the product could be run closer to the upper specification limit.

The maximum economic return from a distillation column can be achieved by minimizing variability and optimizing the quality performance between the distillate and bottoms streams (Figure 7.6). The optimum DNS for a product being sold is usually in the range of 3.0 to 3.5σ. The optimum DNS for a stream that is being recycled back in the process is usually in the range of 1.5 to 2.0σ.

Reduction of variability can be accomplished by the selection of the best control strategy for the distillation column to shed disturbances and by tuning process control loops for

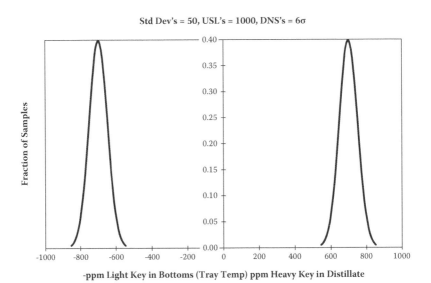

Std Dev's = 50, USL's = 1000, DNS's = 6σ

-ppm Light Key in Bottoms (Tray Temp) ppm Heavy Key in Distillate

**Figure 7.5    Reduction of variability in distillation column.**

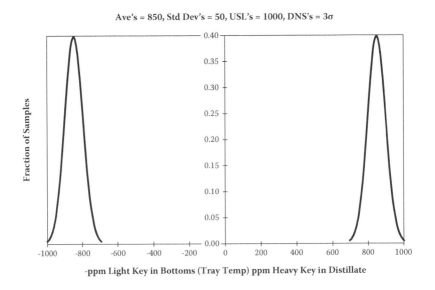

Ave's = 850, Std Dev's = 50, USL's = 1000, DNS's = 3σ

-ppm Light Key in Bottoms (Tray Temp) ppm Heavy Key in Distillate

**Figure 7.6    Optimum economic return from distillation column.**

minimum variability and good response to load changes if needed. If a distillation column is on manual control, then one improvement may be to install a temperature controller for the MRT (most responsive temperature) point in the column to automatically shed many of the disturbances that come to the column. In some cases, a matrix of the response of each variable for the distillation column is measured empirically, and then a dynamic matrix control is used to reduce variability. Online analyzers can be used for direct composition control, but data from production plant distillation columns have shown the MRT (most responsive temperature) point responding 60 to 90 min before the online analyzer responds at one end of the column.

## 7.6  Optimize with Expert System Advisor

The distillate and bottoms streams from a distillation column can be sampled and analyzed once per shift, and then an expert system can be used to recommend adjustments in setpoints for the separation power base and the MRT using a set of backward chaining logic rules. For the case of an upper specification limit of 1,000 ppm heavy key impurity in the distillate stream and a standard deviation of 100 ppm impurity, the desired impurity concentration may be 700 ppm, that is, 3 standard deviations below the upper specification limit. The expert decision rules can be set to make adjustments only if the impurity concentrations are outside of the range from 0.6 to 0.8 times the upper specification limit (USL). For example, the first priority is to check the key impurity separation power. If the key heavy impurity concentration in the distillate is above 0.8 times its upper specification limit and the key light impurity in the bottoms is above 0.8 times its upper specification limit, then increase the separation power base (Rule 1 in Table 7.2). An initial guideline is to increase the separation power base, steam/feed, or reflux/feed by 1%.

**Table 7.2  Backward Chaining Expert Logic Rules**

| Rule No. | Impurity Concentration | | Recommended Action | |
|---|---|---|---|---|
| | Distillate | Bottoms | S/F or R/F | Temp (D/F) |
| 1 | >0.8 Dist USL | >0.8 Bott USL | Incr 1% | None |
| 2 | <0.8 Dist USL | >0.8 Bott USL | None | Incr .3(MRT/stage) |
| 3 | <0.6 Dist USL | >0.6 Bott USL | None | Incr .3(MRT/stage) |
| 4 | >0.8 Dist USL | <0.8 Bott USL | None | Decr .3(MRT/stage) |
| 5 | >0.6 Dist USL | <0.6 Bott USL | None | Decr .3(MRT/stage) |
| 6 | <0.6 Dist USL | <0.6 Bott USL | Decr 1% | None |
| 7 | | | No change necessary | |

Next, balance the quality performance of the distillate stream versus the bottoms stream (Rules 2 through 5 in Table 7.2). If the $D/F$ (distillate/feed) ratio is too low, then raise the MRT (most responsive temperature) setpoint (Rule 2 or Rule 3). An initial guideline is to raise the MRT setpoint by, say, 0.3 times the MRT change/stage shown in Chapter 4, Figure 4.1. The temperature gradient of the MRT was 1.35°C/stage for the C6/C7 system, so raise the temperature setpoint 0.3(1.35) = 0.4°C. If the $D/F$ is too high, then reduce the MRT setpoint by 0.4°C (Rule 4 or Rule 5).

Finally, check to see if the separation power is too high and can be reduced to save energy and increase tower capacity. If the key heavy impurity concentration in the distillate is less than 0.6 times its upper specification limit and the key light impurity in the bottoms is less than 0.6 times its upper specification limit, then reduce the separation power base (Rule 6 in Table 7.2). An initial guideline is to reduce the separation power base, steam/feed, or reflux/feed by 1%.

The trigger points and the size of incremental adjustments can be changed to meet the needs of each distillation column. For the case of a distillate that forms an azeotrope with 30% heavy key impurity, the column may be designed for an upper specification limit of 32% heavy key impurity, and the standard deviation of impurity concentration may be 0.2%. Then, the trigger points for Table 7.2 can be set at, say, 31.6% and 31.2% heavy key impurity to be 2 standard deviations and 4 standard deviations below the upper specification limit.

In some distillation columns, the temperature movement from concentration changes is too small to be measured. For example, an off-spec product with 5 ppm of light key impurity is stripped down to 10 ppb in the bottoms and concentrated to 50 ppm in the distillate. In that case, then, the distillate flow rate can be set to be 10% of the feed rate to get the distillate concentration to increase by 10 from 5 to 50 ppm. The recommended action in Table 7.2 can be to increase or decrease the distillate, or distillate/feed, by 1%.

An expert system will run a set of backward chaining rules in the same order each time the system is run starting with Rule No. 1. If Rule No. 1 is not true, then Rule No. 2 is run. If Rule No. 2 is true, then Rule No. 2 is said to fire. When Rule No. 2 fires, the expert system stops, and the action for Rule No. 2 is recommended. If none of the first six rules fire, then Rule No. 7 fires, and the recommendation is "No Change Necessary."

A situation can arise when operating a distillation column where Rule No. 1 in Table 7.2 can fire to recommend an increase in steam/feed or reflux/feed, but that cannot be accomplished because of a constraint on a manipulated variable. The most simple case occurs when a steam valve or a reflux valve is already 100% open and cannot be increased any further. That is a constraint. In another case, the flowmeter for the steam rate or for the reflux rate is at 100% of span and cannot indicate a higher value for a flow control loop. In yet another case, the vapor traffic going up the column is creating the maximum pressure drop that can be operated successfully through a distillation tray or through a distillation column packing. A higher flow rate of vapor would hold liquid up in the column, and it would be flooded. The column must be kept below flooding. This is another constraint.

**Table 7.3 Distillate Rules When Steam or Reflux is Constrained**

| Rule No. | Impurity in Distillate | Recommended Action Temperature (D/F) |
|---|---|---|
| 1 | >0.8 Dist USL | Decr .3(MRT/stage) |
| 2 | <0.6 Dist USL | Incr .3(MRT/stage) |
| 3 | | No change necessary |

When the separation power is at a maximum constrained value because of boilup or reflux, then Rule No. 1 in Table 7.2 is no longer valid, and a new set of rules is needed. When either boilup or reflux is constrained, then one degree of freedom for control is lost, and only one of the product streams can be kept within specification. A decision would need to be made to keep the distillate within specification or bottoms within specification, but it may not be possible to keep both streams in spec. If the distillate stream is to be kept in spec, then the three rules in Table 7.3 can be used. If the bottoms stream is to be kept in spec, then the three rules in Table 7.4 can be used.

When the operating conditions were optimized for a number of distillation columns, there was an average reduction in energy consumption of 18% valued in the range of $30,000 to $50,000/year, which was also accompanied by

**Table 7.4 Bottoms Rules When Steam or Reflux is Constrained**

| Rule No. | Impurity in Bottoms | Recommended Action Temperature (D/F) |
|---|---|---|
| 1 | >0.8 Bott USL | Incr .3(MRT/stage) |
| 2 | <0.6 Bott USL | Decr .3(MRT/stage) |
| 3 | | No change necessary |

an 18% increase in capacity. One process train involved the distillation of large quantities of water, and the optimization reduced energy consumption by $500,000/year. In another situation, the optimization of distillation operating conditions resulted in the recovery of additional product from the bottoms that was a waste stream going to an incinerator. The value of the recovered product was $220,000/year with no increase in raw material cost or capital. Also, the costs for transporting and incinerating the waste were reduced.

The largest improvements occurred when the product from a distillation column was sold out, and another pound produced meant another pound sold. In one case, the distillation column was being operated at the maximum boilup rate recommended by the vendor of the distillation trays. However, the use of small incremental changes by the sequential optimization routine increased the boilup over a period of several weeks and resulted in the discovery that the trays could be operated at 136% of the maximum capacity recommended by the vendor before flooding occurred. This resulted in additional sales of $800,000 one year during a peak in demand for the product with no added capital. In another case, the product from a new world-scale plant was sold out, and the bottleneck for the plant was the size of the reboiler on a distillation column. The boilup rate for the column was being manipulated by a temperature controller. The capacity of the plant was increased by setting the reboiler to run at its maximum capacity, that is, at the constraint limit, all of the time, and the control strategy was changed to use a temperature control loop to manipulate the reflux flow rate. This resulted in an increase in production that year of $32 million.

Parkinson[3] reported the use of a system called D-POP (Distillation Performance Optimization Program) for optimizing the quality performance from distillation columns via the Internet.

## 7.7 Summary

This chapter described the quality performance of the distillate stream in terms of the average concentration of heavy key impurity and the number of standard deviations from the upper specification limit. The quality performance of the bottoms stream is described in terms of the average concentration of light key impurity and the number of standard deviations from the upper specification limit. The maximum economic return from a distillation column is achieved by reducing variability, striking a balance between the distillate and bottoms quality, and reducing energy consumption.

## Exercises

7.1 If the average impurity concentration is equal to the USL (upper specification limit), how many samples would be out of spec every week assuming 21 samples per week and a normal distribution of error?

7.2 If the light key impurity USL (upper specification limit) for ethanol in water is 0.3% and the standard deviation is 0.025%, what should the target concentration be to achieve a frequency of failure rate of 1 sample out of 790?

7.3 For a binary distillation of benzene and toluene with a USL of 0.2% toluene in the distillate and 1% benzene in the bottoms, how should the MRT setpoint be changed if the distillate contained 0.1% toluene and the bottoms contains 1.5% benzene?

7.4 For a binary distillation of benzene and toluene with a USL of 0.2% toluene in the distillate and 1% benzene in the bottoms, how could the column performance be improved if the toluene in the distillate was 0.05% and the benzene in the bottoms was 0.1%?

7.5 If the upper specification limit for the heavy key impurity in the distillate is 0.5% and the upper spec limit for the light key impurity in the bottoms is 0.2%, which rule would fire in Table 7.2 if the distillate sample analyzed 0.35% heavy key impurity and the bottoms sample analyzed 0.1% light key impurity?

7.6 If Rule No. 4 fired in Table 7.2, should the temperature setpoint be raised, lowered, or left unchanged?

7.7 How large of a temperature change in setpoint would be recommended by Table 7.2 if the temperature gradient for the controlled temperature is 5°C per theoretical stage?

# References

1. Montgomery, D. C., and Runger, G. C., *Applied Statistics and Probability for Engineers*, John Wiley & Sons, New York, 1994.
2. Robinson, J. W., Improve dryer control, *CEP*, 28–33, December 1992.
3. Parkinson, G., Optimizing distillation columns via the Internet, *Chem. Eng.* Vol. 108, No. 11, October 23–25, 2001.

# Chapter 8

# PID Feedback
# Control Loop

This chapter provides an introduction to a feedback process control loop and a description of the action provided by Proportional (P), Integral (I), and Derivative (D).

## Learning Objectives

When you have completed this chapter, you should be able to

1. Understand the concept of the process variable, the controller setpoint, and the error between them
2. Understand the controller action provided by the proportional gain (P)
3. Understand the controller action provided by the integral (I) reset
4. Understand the controller action provided by the derivative (D) rate or preact

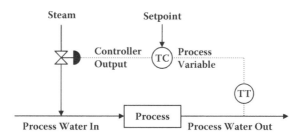

TT = Temperature Transmitter
TC = Temperature Controller

**Figure 8.1   Feedback control loop for heating process water with live steam injection.**

# 8.1  The Feedback Control Loop

Feedback controllers have been used for many years, starting with speed control on windmills and then steam engines[1] and now modern-day cruise control on automobiles. The concept is to increase the flow of energy when the speed falls below the desired value or reduce the flow of energy when the speed is too high.

An example of a feedback control loop is shown in Figure 8.1 for heating the process water with live steam injection. The temperature of the water leaving the process is measured by a temperature sensor, and a transmitter sends the signal to the controller. The desired temperature setpoint is adjusted in the controller, and the difference between the water temperature out of the process and the setpoint is called the *error.* In this example, the process water temperature out of the process is the controlled process variable, that is, controlled to a setpoint. The manipulated variable is the steam flow rate, which is manipulated by the automatic valve that is connected to the controller output.

The controller output is determined by the three controller actions P, I, and D in response to the error and how it changes with time.

## 8.2 Proportional Action

The proportional action of the controller is for instantaneous controller output that is proportional to the error between the process variable and the setpoint in the controller (Equation 8.1):

$$CO = CO_b + K_c\, e \qquad (8.1)$$

where:

$CO$ = controller output
$K$ = proportional gain
$e$, error = process variable − setpoint
subscript b = bias (initial) value
subscript c = controller

If the water temperature out of the process is too hot, the steam valve will be closed in direct proportion to the error. Similarly, if the water temperature out of the process is too cold, the steam valve will be opened in direct proportion to the error. However, if there is a change to increase the temperature setpoint or a change to increase the load (process water flow rate), the proportional action by itself cannot eliminate the error. The new steady-state steam valve position would need to be further open to eliminate error. This is a result of the fact that when the error is zero, the valve would only be open far enough to heat the water to the lower setpoint or the lower flow rate. As a result, the temperature would drop by an amount called offset. The amount of offset can be reduced by increasing the proportional gain in the controller, $K_c$, but the increased gain can cause oscillation and cycling in the process variable.

Some controllers use proportional gain, $K_c$, as the tuning constant. Other controllers use the reciprocal or $100/K_c$ as the proportional band in percent as the tuning constant. A 10% proportional band will change the controller output from 0%

to 100% as the error changes by 10% of the span of the process variable.

The control loop in Figure 8.1 needs to decrease the controller output (steam valve position) as the process variable (temperature) increases, so this loop requires reverse action or negative feedback. For the injection of chilled water to cool the process water, the controller output (cooling water valve position) would need to increase or open if the process variable increased. That would require direct action or positive feedback.

## 8.3 Integral Action

The integral (reset) action in the controller is primarily for eliminating the offset error at steady state (Equation 8.2):

$$CO = CO_b + (K_c/T_i) \int e \; dt \qquad (8.2)$$

where:
$T_i$ = integral (reset) time constant
$t$ = time

In the example shown in Figure 8.1, if the temperature setpoint or process water flow rate (load) increased, then any offset error would be eliminated by the integral action, which would increase the controller output until the error was zero. This automatic reset action essentially moves the valve position or bias to the new position needed at the higher setpoint or higher water flow rate to run at the desired temperature setpoint for the process variable. The integral time constant, $T_i$, is the amount of time to repeat the change in controller output by the same amount the proportional gain changes the controller output for a given error $e$.

Some controllers use $T_i$ as the integral tuning constant in minutes. Other controllers use the reciprocal $1/T_i$ as the reset tuning constant in repeats per minute.

## 8.4 Derivative Action

The derivative (rate or preact) action in a controller is for responding to the rate of change of the error (Equation 8.3).

$$CO = CO_b + (K_c \, T_d) \, de/dt \qquad (8.3)$$

where:

$T_d$ = derivative time constant

The derivative time constant, $T_d$, determines the amount of change in controller output as the rate of change of the error increases or decreases. When the controller output responds directly to the error signal as shown in Equation 8.3, a spike or noisy movement in the process variable can be translated into an immediate movement in the controller output.

## 8.5 Parallel and Series Algorithms

The simplest algorithm for a PID controller is the sum of Equations 8.1, 8.2, and 8.3 as shown in Equation 8.4. This is common in computer-based control systems, and all three control actions are considered to be operating in parallel. However, many industrial analog controllers and microprocessor DCS (distributed control system) controllers use a capacitance lag (filter) of about 0.05 to 0.10 in series with the process variable signal to reduce the effect of derivative action from setpoint changes and from short time constant noise described earlier.[1] When the derivative time constant, $T_d$, is set to zero, there is no difference between the controller action of the parallel and the series algorithm with only P and I action:

$$CO = K_c \, e + (K_c/T_i) \int e \, dt + (K_c \, T_d) \, de/dt \qquad (8.4)$$

## 8.6 Summary

This chapter presented a flow diagram of a PID feedback control loop with the process streams and instrument lines labeled. The proportional, integral, and derivative actions of the controller were defined mathematically and described for the example in the diagram.

## Exercises

8.1 How much does the controller output change if a proportional-only controller gain is 5 and the process variable changes by 5% of its span?

8.2 With a proportional-only temperature controller in Figure 8.1 using a controller gain of 12.5, the process water flow increased and the steam valve position increased by 7% at steady state. What is the offset in outlet temperature in degrees in centigrade if the span of the temperature transmitter is linear from 0°C to 200°C? Was the new steady-state temperature of the process water out too low or too high?

8.3 For a PI controller with a proportional band of 25% and a reset rate of 0.05 repeats per minute, what is the proportional gain of the controller and how many minutes will it take to change the controller output the same amount as the gain for a fixed amount of error?

## Reference

1. Corripio, A. B., *Tuning of Industrial Control Systems*, 2nd ed., ISA, Research Triangle Park, North Carolina, 2001.

## Chapter 9

# Closed-Loop Tuning of Controllers

The tuning of process control loops can have a significant effect on the variability and robustness of the control of a process system. This chapter teaches methods of tuning controllers while they are running in automatic output mode, that is, closed loop. The pattern recognition methods are highly effective.

## Learning Objectives

When you have completed this chapter, you should be able to

1. Know the difference between a self-regulating process response and an integrating process response
2. Understand the procedures and methods for tuning controllers for self-regulating and integrating process responses while the controllers are running in automatic output mode
3. Recognize the desired response pattern when a step change in setpoint is introduced to the controller for a self-regulating process variable with optimum

proportional gain and insignificant integral (reset) action. The height between the first peak and valley will be 16% of the height of the step change in controller setpoint

4. Measure the response time for a self-regulating process variable to reach the first peak after a step-up change in setpoint when tuning a self-regulating process response

5. Recognize the response pattern of an integrating process variable, such as distillation column bottom level or reflux drum level, when a step change in setpoint is introduced to the controller with a proportional gain of 1.25 and an integral time constant of 30 min

6. Measure the response time for a liquid level process variable to reach 95% of the step change in setpoint when tuning a liquid level control loop for a reflux drum or distillation column base

The process dynamics of most process variables can be characterized as a *self-regulating process response.* One example is the response of a liquid flow rate when a valve position is opened. The liquid flow rate will increase from the initial flow rate to a new steady-state flow rate. Another example is the response of the temperature of a liquid flowing through a heat exchanger that is heated with steam. When the steam valve position is increased, the temperature of the liquid outlet will increase to a new steady-state temperature.

The process dynamics of a few process variables can be characterized as an *integrating process response.* One example is the response of the liquid level in the bottom of a distillation column when liquid is being pumped out with a centrifugal pump and the liquid flow rate is restricted by an automatic valve after the pump. The column base liquid level may be steady with an initial automatic valve position, and opening the valve to a new position will increase the liquid flow rate. The difference between the two flow rates will be integrated over time as seen by the response in liquid level going down. Eventually, the column base liquid level will go to zero without further intervention. In other words, the liquid level response is integrating and is not self-regulating.

## 9.1 Trial-and-Error Tuning of Control Loops

Many process control loops are tuned by trial and error while the controller is in automatic output mode. Everyone who has ever tuned a process control loop has used trial and error at one time or another. Some people have learned the art of tuning process control loops by trial and error and have become highly proficient by learning from their own experience or from the experience of a mentor or a network of practitioners. Trial-and-error tuning consists of changing one or more of the PID (proportional, integral, and derivative) actions in the controller by some amount and observing the response of the control loop to see if it changes for the better or worse. The proportional gain may be changed by 10% or 20% or may be doubled or cut in half. The integral and derivative may also be changed by the same amount or turned off.

One good point about trial and error is that the controller is in automatic, and the controller settings can be returned back to the initial values if they are recorded somewhere or if they can be remembered. Another good point about trial-and-error tuning is that the usual equipment existing in a control room for recording a process variable response with time can be used without any additional tools or computer programs. One concern about trial-and-error tuning is that a beginner may not recognize whether the change in process variable response is better or worse than the response was with the initial values. Another concern is that trial-and-error experimentation may involve a long learning curve. Learning how to tune a control loop may involve many setpoint changes and process upsets. Also, trial-and-error tuning can sometimes lead to mindless tweaking of the controller settings depending on the individual and the plant circumstances.

The two main types of disturbances introduced into a controller for closed loop tuning are:

1. A step change in controller setpoint
2. A step change in load

A step change in load can be simulated by first placing a controller in manual output; second, introducing a step change in the controller output (valve position); and third, placing the controller back in automatic.

## 9.2 Ultimate Gain Tuning of Control Loops

The Ziegler–Nichols[1] closed-loop tuning method was published in 1942 and has become a classic method of tuning control loops. Many suppliers of controllers provide the Ziegler–Nichols method of tuning in the user guide for their controllers. The method consists of first turning off the integral (reset) and derivative (rate) action in the controller and, second, finding the ultimate proportional gain (UG) that will just sustain continuous cycling of the process variable and measuring the ultimate peak-to-peak time period (UTP). These two values (UG and UTP) characterize the closed-loop response.

When only *P* and *I* (proportional and integral) actions are used in a controller, the recommended *P* (proportional gain) is a fraction of the ultimate gain, 0.45 UG. Some controllers use proportional band, which is 100 divided by the proportional gain.

The value of *I* (integral action) is set by $T_i$ (integral time constant) in minutes in some controllers, or the reciprocal, $1/T_i$, as reset rate in repeats/minute, in other controllers. Ziegler and Nichols recommended the value for integral time constant to be set as a fraction of the ultimate peak-to-peak time period, $T_i = 0.83$ UTP, in units such as minutes. The controller action from the proportional gain times error will be repeated in the length of time set by the integral time constant.

When all three PID actions are used, the recommended proportional gain is increased to 0.60 UG, the integral time constant is reduced to 0.50 UTP, and the derivative time constant is set to 0.125 UTP. In other words, the addition of derivative action allows the use of a higher proportional gain and a shorter integral (reset) time constant. Tuning a controller to these constants will give an aggressive response by the controller to shed disturbances to a control loop.

Robbins[2,3] recommended using a proportional gain of 0.3 UG for minimum variability in the process response. The recommended integral time constant is 1.0 times UTP for good response to load change and 2.0 times UTP for minimum IAE (integral of absolute error) in response to a step change in setpoint.

One good feature of the Ziegler–Nichols closed-loop method is that it can be learned more quickly than starting with trial and error alone. There is a procedure to be followed, and the pattern of sustained cycling is easy to recognize. The Ziegler–Nichols method is often completely acceptable for tuning control loops that respond quickly, for example, liquid flow rate control loops that respond with an ultimate peak-to-peak time period (UTP) of 5 to 15 s.

One concern with the ultimate gain method is that the process must be run at a high proportional gain that will sustain a steady-state cycle, that is, ringing, in the control loop. Any higher proportional gain in the controller may cause the control loop oscillations to grow to a higher amplitude and become unstable. Another concern of this ultimate gain method is that a long time may be required to wait for two or three cycles to develop a pattern of oscillation to see if it is damped, sustained, or growing. If the ultimate peak-to-peak time period is 1.0 to 1.5 h, the time required may be 4.5 h or more for one test. Also, some control loops may have no dead time or may have an integrating response that may not sustain a cycle. These concerns provided the motivation for developing the pattern recognition methods for closed-loop tuning in Section 9.5.

# 9.3 Troubleshooting an Oscillating Control Loop

Sometimes an existing process control loop is already running with sustained cycling, and tuning of the control loop can be approached as a troubleshooting activity. There can be a number of possible causes for the process variable oscillating up and down, and the problem is not always due to poor tuning. If the controller is put in manual output mode so that it does not respond to feedback from the process variable and the oscillations continue, the problem is most likely due to causes other than the tuning of that one controller. One possibility is interaction from another control loop that causes the process variable to cycle. If a controller is put in manual mode and the oscillations stop, the problem may possibly be solved with controller tuning.

One of the first patterns to observe in a cycling control loop is the timing of the peaks and valleys in the process variable (controller input) and the manipulated variable (controller output) and whether or not the cycles are in phase with each other. Observe the point in time when the controller output is at a peak or a valley. If the process variable is also at a peak or a valley at the same time, the problem may be due to the proportional gain being too high, that is, too narrow a proportional band. However, if the process variable reaches a peak or a valley at a different point in time than the controller output, the problem may be due to the integral (reset) time constant being too short relative to the peak-to-peak time period. The integral time constant needs to be about the same as the peak-to-peak time period. Reducing the integral (reset) time constant in the controller to be shorter than the peak-to-peak time period usually does not speed up the control response.

The recommendation for tuning is to set the integral time constant equal to the peak-to-peak time period and increase the proportional gain by the same factor that the integral time constant was increased. Increasing the proportional gain in the controller will usually shorten the response time.

If a process variable, such as a flow rate, is oscillating with a square wave pattern and the controller output looks like a zigzag sawtooth shape, the cause may be due to a valve problem such as sticking or friction. Another problem may be that the controller output signal is hitting a saturation point, for example, 0% or 100% open. There can also be problems with hysteresis if there is no positioner on the control valve. When there is hysteresis, the valve position may be low if the control air pressure signal is going up, but the valve position may be high if the control signal is going down. In these cases, the problem may be due to mechanical reasons and not controller tuning. In general, when the process variable is oscillating but the shape of the response is not sinusoidal, there may be a mechanical problem.

## 9.4 Quarter Decay Ratio Tuning of Control Loops

When controller constants are used from the Ziegler–Nichols closed-loop tuning method, the response pattern of the process variable tends to give a one-quarter amplitude decay ratio, or quarter decay ratio (QDR). In other words, when a step change in setpoint is introduced into the controller, the first peak (overshoot) in the process variable response will be four times the height of the second peak (overshoot). The first peak may overshoot the setpoint by 50% of the step change in setpoint, the first valley may undershoot the setpoint by 25%, and the second peak may overshoot the setpoint by 12.5% of the height of the step change in setpoint. The first overshoot of 50% is 4 times higher than the second overshoot of 12.5%.

The height between the first peak and valley, that is, 50% plus 25%, is 75% of the height of the step change in setpoint. This response pattern can be recognized as being aggressive. There is a lot of control action in response to a disturbance to the control loop.

Generally, when the proportional gain in a controller is less than the value recommended by the Ziegler–Nichols method, the integral (reset) time constant can be reduced until a quarter amplitude decay response is developed. However, a low proportional gain and a short integral time constant can give a process response that is very sluggish because of the low gain. In that case, the integral time constant can be increased to be equal to the peak-to-peak time period, and the proportional gain can be increased by the same factor that the integral time constant was increased.

## 9.5 Pattern Recognition Tuning of Self-Regulating Control Loops

Robbins[2,3] reported the development of a closed-loop tuning methodology that does not require any of the loops to be run with sustained cycling. The steps for the Robbins tuning method for self-regulating control loops are as follows:

1. Record the existing settings in the controller.
2. Turn off the reset, or set the integral time constant at least four times the response time.
3. Turn off the derivative action, that is, $D = 0$.
4. Introduce a step change in the controller setpoint, and observe the height between the first peak and valley in the process variable response. The optimum proportional gain will give a height between the first peak and valley that is 16% of the size of the step change.
5. Record the response time for the process variable to reach the first peak after introducing a step up change in setpoint.

6. Set the integral (reset) time constant in the controller to be equal to the response time.

This will give an excellent response to load changes. Generally, no derivative action is desired for distillation controllers, because any noise in the signal for the controlled process variable can be translated into controller output and valve movement. Smooth controller output to a control valve is usually desired for distillation. When derivative action is used, the recommended setting is 0.10 to 0.15 times the response time.

The recommended method for tuning level control loops is given in Section 9.8. It is different from the method for tuning controllers with a self-regulating process variable response.

During start-up of a new distillation column, one question can be asked about which controllers should be tuned first and which ones should be tuned last. First, the liquid level control loops for the reflux drum and the column base liquid level can be set with the starting tuning constants recommended in Section 9.8, that is, a proportional gain of 1.25 and an integral time constant of 30 min. Next, the fast-responding liquid flow rate control loops should be tuned. When a cascade (master-slave) temperature controller is used, the slow-responding temperature control loop is the master that manipulates the setpoint to a fast-responding flow rate slave control loop. The master temperature controller can be placed in manual output mode to introduce a step change in setpoint to the flow rate slave controller. The slave control loop needs to be tuned before the master control loop is tuned.

One of the guidelines for Step 4, when there is no peak in the process variable after a step change up to a higher setpoint, is to double the proportional gain. Of course, if the step change is down to a lower setpoint, the response time would be the time to reach the first valley in the process variable response. Once the height of the peak-to-valley response is in the range of 5% to 50% of the size of the step change, then the proportional gain can be changed by an increment of about 20% of the gain. Often, the response is not linear, and a smaller proportional gain is required to get the 16% peak-to-valley response in one step direction compared to the other, that is, a step-up versus a step-down. In that case, use the smaller proportional gain and the longer response time.

Another guideline is that an integral time constant of 4 times the response time will minimize the effect from the integral (reset) action, as indicated in Step 2. Yet another guideline is that an integral time constant significantly less than the response time will generally destabilize the control loop and give variability with a long peak-to-peak time period.

The development of the Robbins[2,3] closed-loop tuning methodology was the result of about 11 years of development with collaboration and feedback of comments from others. A number of different methods for tuning process control loops were tested along the way, and this new closed-loop method using pattern recognition was determined to be the most cost-effective.

Dynamic computer simulations of control loops were run with thousands of combinations of process models and hundreds of production plant process control loops were tuned for real plant experience. Most of the plant controllers were on distillation columns, but control of reactors, furnaces, dryers, extruders, and other unit operations was also improved by tuning with these techniques. Most of the computer simulations used a dead time and two capacitance lags, and many different combinations of those three process parameters were studied. Initially, the control loop response was tuned for minimum IAE (integral of absolute error), that is, minimum deviation of the controlled process variable (controller input) from the setpoint, after a step change in setpoint was made in the controller. A pattern was recognized: the minimum IAE occurred when the overshoot-to-undershoot ratio was about seven to one. In other words, after a step change of 100%, the first overshoot was one seventh of that, that is, 14% of the step change, and the first undershoot was one seventh of the overshoot, that is, 2% of the step change. This gave an amplitude decay ratio of about 1/49; that is, the first overshoot was about 49 times higher than the second overshoot. It was important to develop the optimum 7 to 1 pattern of response without any significant influence from integral (reset) action or derivative action in the controller. When there is no automatic reset, that is, no integral action in the controller, the first peak in the process variable response may not get up to the new setpoint, and the control loop can be left with offset from the setpoint. So, a method was developed that looked for the height between the first peak and valley in the process variable response to be 16% of the height of the step change. In other words, for a step change in setpoint up 10°C, the height between the first peak and valley is 1.6°C when the proportional gain is at the optimum value. Similarly, when the step change in setpoint is down 10°C, the valley will be first, and the height between the first valley and peak is 1.6°C. The size of step change in setpoint in an actual controller needs to be small enough so the manipulated variable (controller output) does not saturate at 100% or 0%, because that can interfere with the development of the height between the peak and valley and with the response time.

Figure 9.1 shows the pattern that can be recognized when a control loop is tuned for minimum IAE (integral of absolute error) after a setpoint change. The setpoint for the process variable was increased by a step change from 50 to 60 at time zero. The process variable crossed the new setpoint at 10 min after the setpoint change and reached the first peak at 13.6 min. The time to reach the first peak is called the response time in the descriptions that follow. The first peak height in the process variable response was 61.6, and the first valley height was 59.6; so the height between the first peak and valley was 20% of the size of the step change. The decay ratio was 400; that is, the first peak was 61.600, and the second peak was 60.004. The controller output, that is, the manipulated variable, stayed within the range of 0% and 100%. The parameters used for the simulation were process gain ($K_p$) of

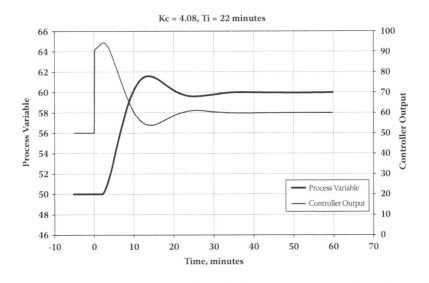

**Figure 9.1 Control loop response with optimum gain.**

1.0, a dead time (DT) of 2 min, first capacitance lag ($T_1$) of 20 min, and second capacitance lag ($T_2$) of 2 min. The tuning constants consisted of a controller gain, $K_c$, of 4.08 and an integral time constant, $T_i$, of 22 min.

## 9.6 Effect of Proportional Gain

When the proportional gain, $K_c$, was half of the optimum, the controlled process variable did not show any overshoot, that is, no peak or valley, in process variable response to a setpoint change (Figure 9.2). The process variable took 20 min to move 95% of the step change. When the control objective is to tune for no overshoot, use the same first four steps, as shown in Section 9.5. Then, set the proportional gain in the range of 0.5 to 0.6 times the optimum gain, and set the integral time constant to be 2 times the response time found in Step 5.

When the controller gain, $K_c$, was one fourth of the optimum gain, the process variable response was sluggish and reached 95% of the step change in 60 min (Figure 9.3). The

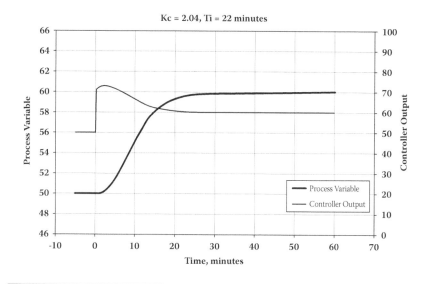

**Figure 9.2    Response with 0.5 times optimum gain.**

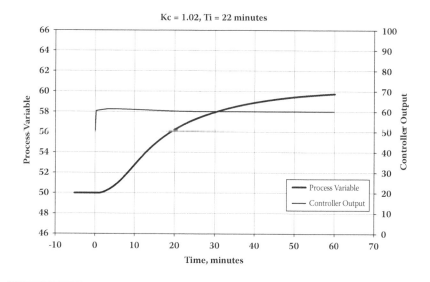

**Figure 9.3    Response with 0.25 times optimum gain.**

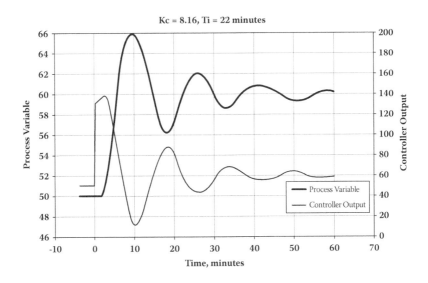

**Figure 9.4    Response with 2 times optimum gain.**

controller output moved to the new valve position quickly and remained there. This is almost identical to the open-loop response when the controller is placed in manual output mode, and a step change in controller output (valve position) is introduced.

When the controller gain, $K_c$, was twice the optimum value, the height between the first peak and valley was 94% of the size of the setpoint change (Figure 9.4). The amplitude decay ratio for this response is 2.9, that is, the first peak overshot the setpoint by 5.8, and the second peak overshot by 2.0. The controller output peaked at nearly 140% in this simulation. An actual controller would have been constrained by a maximum value of 100%, so the shape of the response would have been affected. A smaller setpoint change should be used to avoid the controller output reaching saturation at either 100% or 0%.

The control loop will sustain cycling, that is, ringing, when the proportional gain in the controller, $K_c$, is about 3.3 times the optimum gain (Figure 9.5). This is close to the ultimate gain and ultimate peak-to-peak time period used in the Ziegler–Nichols closed-loop tuning method. The peak-to-

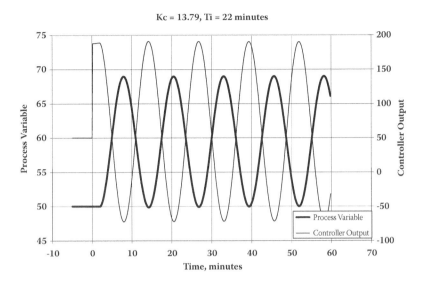

**Figure 9.5 Response with 3.3 times optimum gain.**

peak time is nearly the same as the response time of 13 min in Figure 9.1. The height between the first peak and valley is about 200% of the size of the step change in setpoint. One point of interest is that when cycling is due to a high proportional gain, the controller output and process variable are in phase; that is, the peaks and valleys occur at the same point in time in the cycle. An actual controller would have a constrained controller output between 0% and 100%, so the shape of the process variable response may not be sinusoidal as shown in Figure 9.5.

# 9.7 Effect of Integral (Reset) Action

A control loop will also sustain continuous cycling when the integral time constant is too short, that is, about 0.3 times the response time to reach the first peak (Figure 9.6). However, it is significant that the process variable and controller output cycles are out of phase with each other. When the controller output is at a peak or valley, the process variable is near its setpoint.

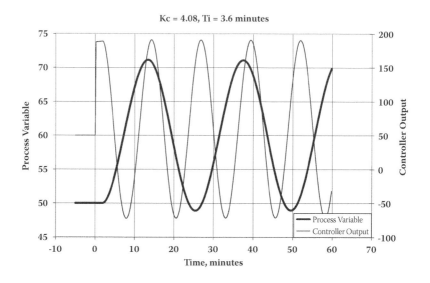

**Figure 9.6   Response with $T_i$ = 0.3 times response time.**

Once again, sustained cycling occurs when the peak to valley is 200% of the size of the step change in setpoint. However, note that the peak-to-peak time period in the process variable is 24 min, that is, about twice as long, when the cycling occurs with too much reset. That is because the proportional gain, $K_c$, is lower than it was in Figure 9.5. A control loop responds faster with high gain and slower with low gain.

The optimum proportional gain, $K_c$, gave a height between the first peak and valley that was 16% of the size of step change in setpoint (Figure 9.1). The integral time constant, $T_i$, was about 2 times the response time of 13 min to reach the first peak for minimum variability, that is, IAE, from a setpoint change. However, this gives sluggish response to a change in load (Figure 9.7).

One example of a load change is to suddenly reduce the feed rate of water to the process shown in Chapter 8, Figure 8.1. When the water feed rate is reduced, the temperature goes up and the controller output or steam valve position has to be decreased by the integral action of the controller to keep the outlet temperature at the setpoint of 54. The response

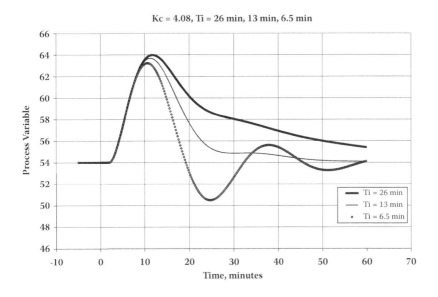

**Figure 9.7   Control loop response to load change.**

to a load change is excellent when the integral time constant is set equal to the response time, that is, 13 min, observed from a setpoint change. Reducing the integral time constant to half the response time, that is, 6.5 min, caused instability in the control loop and high variability. The response time to reach the first peak after a load change shown in Figure 9.7 was 11.6 min, which is nearly the same as the response time to reach the first peak after a change in setpoint, that is, 13 min, as shown in Figure 9.1. Reducing the integral time constant, that is, increasing the automatic reset rate, did not have a significant effect on the response time to reach the first peak.

Reducing the integral time constant from 2 times the response time to 1 times the response time has a dramatic effect on improving the response to load changes in Figure 9.7. The cost of this change is that the variability from a setpoint change is no longer at the minimum, as shown in Figure 9.8. The height between the peak and valley in Figure 9.8 was about 36% of the height of the step change when the integral time constant was 13 min. This trade-off can be beneficial if

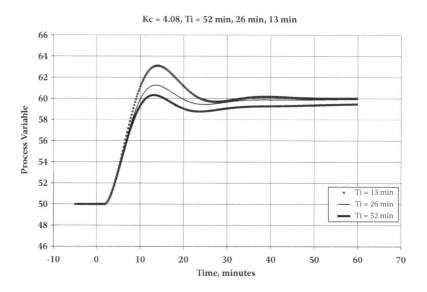

**Figure 9.8    Control loop response to setpoint change.**

the control system is required to shed large load disturbances. Starting up a distillation column by ramping up the feed rate is another example of a load change. The steam flow rate to the reboiler and the reflux and distillate flow rates would need to change in response to the change in load.

The height between the first peak and valley can be reduced by decreasing the proportional gain to the range of 0.5 to 0.6 times the optimum gain, as shown in Figure 9.2.

The first step in tuning a process control loop is to record the existing settings, so they can be restored if desired. Next, introduce a step change in setpoint to the controller, and observe if there is a peak and valley in the response. If there is no peak, the proportional gain can be doubled, and another step change can be introduced into the controller, for example, a step back down to the original setpoint. If the controller has been tuned for quarter amplitude decay and the integral time constant is shorter than the response time, then increase the integral time constant to be equal to the response time. Then increase the proportional gain by the same factor.

The integral time constant can be adjusted to about 4 times longer than the response time, and this will reduce any effect from the integral (automatic reset) action. The response time is the time that elapses between the introduction of a step change in setpoint and the first peak in the response of the process variable. With $T_i$ = 52 min, the response time was 13.2 min for the process variable to reach a peak height of 60.3 and a valley height of 58.7 in Figure 9.8. Since the reset was so slow, the first peak did not overshoot the setpoint very far, and there was a significant offset between the process variable and the setpoint for a long period of time.

Next, the integral time constant, $T_i$, can be set equal to the response time for excellent response to a load change or twice the response time for minimum variability from a setpoint change.

The value of $T_i$ can be set to about 2 times the response time of 13.0 min, that is, 26 min, for minimum variability to a setpoint change. For many combinations of $K_p$, $K_c$, DT, $T_1$, and $T_2$ for self-regulating control loops, a correlation was observed for minimum IAE, as shown in Equation 9.1. In this case, the optimum value of $K_c$ was found to be 4.08, and $K_p$ was 1.0, so the calculated $T_i$, was 1.75 times the response time of 13.0 min, that is, 22.7 min.

$$T_i = [0.24 + (0.37 \ K_c \ K_p)] \text{ response time} \qquad (9.1)$$

where:

$T_i$ = integral time constant for controller (in minutes) for minimum IAE for setpoint changes

$K_c$ = controller proportional gain

$K_p$ = process gain

Response time = elapsed time between step change in setpoint and first peak (in minutes)

The self-regulating process gain, $K_p$, can be measured from the setpoint change and controller output (valve position) change (Equation 9.2).

$$K_p = \frac{(\text{set point change, \% of span})}{(\text{controller output change, \% at steady state})} \qquad (9.2)$$

For example, suppose the step change in the controller was from 70°C to 76°C with a 200°C span, and the steady-state valve position increased from 50% to 52% open. The process gain, $K_p$, would be $[(76 - 70)/200]/[(52 - 50)/100] = 1.5$.

Tuning a control loop for minimum IAE after a step change in setpoint led to the discovery of the 7/1 overshoot/undershoot ratio for finding the optimum proportional gain, $K_c$. However, the integral time constant is too long; that is, the reset is too slow, for good response to load changes. An integral time constant of 26 min did not give a good response to a load change, as shown in Figure 9.7.

The optimum integral (reset) time constant for excellent response to a load change is about 1.0 times the response time, that is, 13 min. Reducing the integral time constant, $T_i$, to 6.5 created instability in Figure 9.7. A correlation for many combinations of self-regulating control loops is shown in Equation 9.3. In other words, with $K_c = 4.08$ and $K_p = 1.0$, the optimum integral time constant for excellent response to load change was calculated as 0.88 times the response time, or 11 min.

$$T_i = 0.55 \, (K_c \, K_p)^{0.33} \text{ response time} \qquad (9.3)$$

where:

$T_i$ = integral time constant (in minutes) for excellent response to load changes

The typical response times for variables in distillation column control are as follows:

- Feed, reflux, or steam flow rate controller, 0.1 to 0.4 min
- Column pressure controller, 5 to 15 min
- Reflux drum or column base level controller, 10 to 30 min
- MRT (most responsive temperature) controller on a distillation column, 10 to 120 min

## 9.8 Pattern Recognition Tuning of Integrating Control Loops

The level controllers for the reflux drum and the column base level for a distillation column are examples of integrating process response. If a level controller is put in manual output mode and the valve position for the liquid flow out of a reflux drum is increased, the level in the drum will integrate the difference, that is, sum the difference over time, and eventually empty the drum. Similarly, decreasing the valve position will allow the drum to fill and eventually overflow. Consequently, the Robbins method of pattern recognition for tuning of level controllers uses a different set of rules as follows:

1. Record the existing settings in the controller.
2. Turn off the reset, or set the integral time constant to 30 min or longer.
3. Turn off the derivative action, that is, $D = 0$.
4. Set the proportional gain to 1.25.
5. Introduce a step change in the controller setpoint.
6. Record the response time for the control loop to reach about 95% of the step change in setpoint after introducing the change.

7. Set the integral (reset) time constant in the controller to be equal to the response time.

A reflux drum is sized and purchased to retain some inventory of liquid and dampen the response in the flow rates leaving the drum, namely, the reflux and distillate. So, there is no financial incentive for controlling the liquid level within a narrow band. On the contrary, there is a penalty for tight level control by moving the flow rates to overshoot the final flow rate and create high variability. Another issue is that a level controller does not actually need any integral (reset) action. A proportional-only controller could be used with a proportional gain, $K_c$, of 1.25 and a controller output bias of 50%, so that the outlet liquid valve could be wide open when the drum was 90% full and completely closed when the drum was 10% full. This is equivalent to a proportional band of 80%, that is, $100\%/K_c$. A proportional gain, $K_c$, of 2.0 could be used so that the valve would be wide open when the drum was 75% full and completely closed when the drum was 25% full.

The dead time for a level control loop is usually very small, so the loop has very little tendency to cycle until the proportional gain is quite high. As a result of all these considerations, the recommended procedure for tuning the level controller for the reflux drum is to simply use a proportional gain in the controller of 1.25 to 2.00 and start with an integral time constant, $T_i$, of 30 min. Introduce a step change in level setpoint, and observe the response time reach 95% of the step change. The level may only approach the new setpoint and may not overshoot as shown in Figure 9.9 for a constant liquid flow rate into the drum of 40% of the maximum flow rate and a step change in level setpoint from 50% to 60%. Then set the integral time constant equal to the response time. The response to a load change, that is, a step change in liquid flow rate into a reflux drum from 40% to 70%, is shown in a simulation in Figure 9.10.

Tuning the control loop for the liquid level in the bottom of a distillation column can be similar to a reflux drum. However,

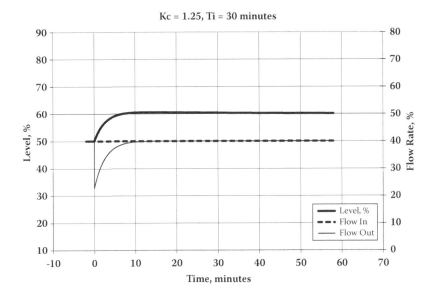

**Figure 9.9    Level control response to setpoint change.**

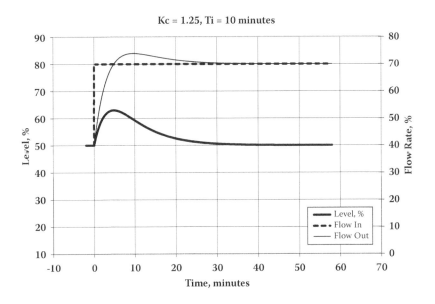

**Figure 9.10    Level response to load change.**

the proportional gain may need to be in the range of 2 or higher if a thermosiphon reboiler is used that requires a narrow range of liquid level for good operation.

## 9.9 Summary

This chapter described the concept of a self-regulating process response and an integrating process response. Several methods were described for tuning a control loop while the controller is in automatic output mode. These include the trial-and-error method, the Ziegler–Nichols ultimate gain method, and the Robbins pattern recognition methods.

## Exercises

9.1 With a controller in automatic output, a proportional gain of 1.0, an integral time constant of 3.0 min, and a derivative of zero, the response to a step change in setpoint gave a quarter amplitude decay, and it took 39 min to reach the first peak. What is the first indication that the controller could be tuned better? What changes would be recommended for the tuning constants?

9.2 When a step change was introduced to a controller setpoint, the process response reached the first peak in 5 min, the proportional gain was 2.5, the integral time constant was 20 min, the derivative was zero, and the height between the first peak and valley was 95% of the height of the step change. Should the integral time constant be increased or decreased? Should the proportional gain be increased or decreased?

9.3 When the setpoint is changed 10% of span, the control valve moves from 60% to 100% controller output and stays there for 10 min while the process variable (temperature) rises to its first peak in 15 min. The proportional band is 15%, the integral time is 45 min, and the derivative is zero. What is the next recommended step in tuning the control loop in automatic output mode?

# References

1. Ziegler, J. G., and Nichols, N. B., Optimum settings for automatic controllers, *Trans. ASME*, 64, 759, November 1942.
2. Robbins, L. A., Tune control loops for minimum variability, *CEP*, January 2002.
3. Robbins, L. A., Setpoint tuning method gives excellent response to load changes, *Hydrocarbon Proc.*, September 2002.

# Chapter 10

# Open-Loop Testing of Process Response

This chapter describes how to characterize the response of a process variable with a controller in manual output mode, that is, with no feedback from changes in the process variable. Then tuning rules based on mathematical models of the response are used to estimate tuning constants for the controller.

## Learning Objectives

When you have completed this chapter, you should be able to

1. Understand some procedures for testing the open-loop response of processes for characterization and mathematical modeling

2. Calculate controller tuning constants by several different tuning rules

## 10.1 Self-Regulating Process Response

As discussed in Chapter 9, the process dynamics of most process variables can be characterized as a self-regulating process response. When the controller output changes the automatic valve position of the manipulated variable, the process variable moves to a new steady-state value.

The most common model of a self-regulating process response is a first-order plus dead time (FOPDT) response. The dead time response can be conceptualized as plug flow through a length of pipe where there is a transport delay time before a process variable change appears abruptly at the end of the pipe. A first-order process response can be conceptualized as a perfectly mixed continuous stirred tank reactor (CSTR) with a fixed volume and a fixed flow rate through the tank to give a fixed residence time. This can also be called a first-order capacitance lag. When a process variable step change is introduced to the inlet of a CSTR, the outlet change follows an exponential decay. Initially, the outlet variable changes rapidly, but the rate of change decays as the new steady-state condition is approached.

The process flow diagram in Figure 10.1 shows the concept of an open-loop process response that includes a dead time plus two first-order capacitance lags. The process water that is heated with live steam injection moves in plug flow through a

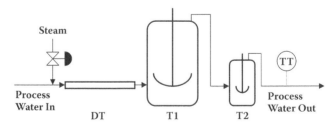

**Figure 10.1   Process concept for modeling dead time plus two first-order capacitance lags.**

length of pipe followed by two CSTRs with different volumetric capacities. In this case, a step change can be introduced to increase the position of the steam valve, and first there will be a transport time delay for the higher-temperature water to be transported through the pipe. The temperature rise will appear abruptly at the end of the pipe and be modeled as a dead time (DT). DT is equal to the volume of the pipeline divided by the volumetric flow rate of the process water. Next, the higher-temperature water will enter the first CSTR with a residence time $T_1$, and the outlet from the first CSTR enters the second CSTR. The capacitance lag times, $T_1$ and $T_2$, are equal to the volume of the CSTR divided by the volumetric flow rate of the process water. The second capacitance lag time, $T_2$, can model the combined response lag of thermowells, sensors, and the movement of automatic valves. The capacitance lag times, $T_1$ and $T_2$, can also be referred to as first-order filter constants.

A computer simulation of a process with a dead time, DT, of 2 min and a first-order capacitance lag, $T_1$, of 22 min gave an open-loop response shown in Figure 10.2. This was compared with an open-loop response with DT = 2, $T_1$ = 20, and $T_2$ = 2 min. Adding the second capacitance lag, $T_2$, caused the process variable response to be rounded a little after the dead time, which is typical of a plant process response.

## 10.2 Ziegler–Nichols (Z–N) Open-Loop Tuning Rules of Thumb

Ziegler and Nichols[1] presented an open-loop test method in their classic paper on tuning control loops. The method consists of starting with the control loop at steady state, putting the controller in manual output so that there is no feedback response from any change in the process variable, and introducing a step change in the controller output (valve position). Then the results are used to characterize the process response by an apparent

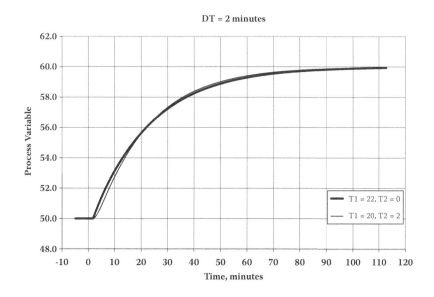

**Figure 10.2 Open-loop process response to step change from 50% to 60% controller output.**

dead time and the maximum rate of change in the process variable, that is, the maximum process response rate.

Two data points were taken from each response curve in
Figure 10.2: when the process variable had reached 10% and
20% of the steady-state response. The results are listed in
Table 10.1. These data can be used to calculate the maximum
process response rate and apparent dead time for each case.

**Table 10.1  Data Points from
Figure 10.2 for Z-N Calculations
with Controller Output Step from
50% to 60%**

| DT | $T_1$ | $T_2$ | Time (min) | Process Variable (%) |
|----|-------|-------|------------|----------------------|
| 2  | 22    | 0     | 4.1        | 51                   |
| 2  | 22    | 0     | 6.7        | 52                   |
| 2  | 20    | 2     | 5.5        | 51                   |
| 2  | 20    | 2     | 8.1        | 52                   |

The open-loop process gain, $K_p$, for a self-regulating process response is calculated from Equation 10.1.

$$K_p = \frac{\%\ \text{change in Process Variable at steady state}}{\%\ \text{change in Controller Output}} \qquad (10.1)$$

The process variable moved from 50% to 60% at steady state for an open-loop step change in controller output from 50% to 60%, so $K_p = 1.0$.

The maximum process response rate, MPRR, is calculated from Equation 10.2.

$$\text{MPRR} = \frac{\%\ \text{change in Process Variable}}{\text{Time, minutes}} \qquad (10.2)$$

For DT = 2, $T_1 = 20$, and $T_2 = 2$ and a step change in controller output of 10% in Table 10.1, the maximum process response rate was calculated as (52 − 51%)/(8.1 − 5.5 min) = 1/2.6 = 0.385% change in process variable/minute using Equation 10.2.

The apparent dead time, ADT, is determined by extrapolating the maximum process response rate, MPRR, back to the starting value of the process variable. The maximum process variable response from 51% to 52% took 2.6 min. A straight line extrapolation back another 2.6 min would get back to the starting point of the process variable of 50%. So, 5.5 − 2.6 = 2.9 min is the apparent dead time, ADT.

The recommended proportional gain is 0.9 times the step size in controller output divided by the maximum response rate of process variable and the apparent dead time (Table 10.2). The CO step size was 10% as listed in the title of Table 10.1. So, the recommended controller gain would be $K_c = (0.9)(10)/(0.385)(2.9) = 8.1$. The Z-N open-loop tuning rules of thumb recommend setting the integral (reset) time constant to 3.33 times the apparent dead time (Table 10.2). In this case,

**Table 10.2 Formulas for Ziegler–Nichols Open-Loop Tuning Rules for PI Controller**

| Tuning Constant | Formula |
|---|---|
| $K_c$ | $\dfrac{(0.9 \text{ CO step size})}{(\text{MPRR ADT})}$ |
| $T_i$ | 3.33 ADT |

the recommended integral time constant, $T_i$, would be (3.33) (2.9) = 9.7 min.

The Z-N equations for predicting controller tuning constants from open-loop test results give a recommended proportional gain of $K_c$ = 12.3 and integral time constant of $T_i$ = 6.3 min. These are very aggressive tuning constants that would give oscillations and high variability compared to the optimum tuning constants of $K_c$ = 4.08 and $T_i$ = 13 min that were found by the pattern recognition method in Chapter 9, Section 9.5, for the same process.

The example in Figure 10.2 required 112 min (dead time plus 5 times the capacitance lag times) for the process variable to move 99% of the way to the new steady-state value. This time or more is needed for determining the steady-state process gain, $K_p$. By comparison, the time to reach the first peak and valley in a closed-loop test took 25 to 30 min in Chapter 9, Figure 9.1, for the same process.

## 10.3 Two-Point Characterization of FOPDT Process Response

Corripio[2] reviewed the two-point Smith method for characterizing the open-loop process variable response to a step change in controller output by an apparent dead time, ADT, and an apparent first-order time constant, TFO. The two data points on the process variable response curve are taken at

T28.3 and T63.2. The data taken are the time elapsed after the step change in controller output for the process variable to reach 28.3% and 63.2% of the steady-state change. The ADT and TFO are calculated from Equations 10.3 and 10.4:

$$\text{TFO} = 1.5 \, (\text{T63.2} - \text{T28.3}) \qquad (10.3)$$

where:

    TFO = first-order time constant (min)

    T28.3 = time for PV to reach 28.3% of change to steady
        state

    T63.2 = time for PV to reach 63.2% of change to steady
        state

$$\text{ADT} = \text{T63.2} - \text{TFO} \qquad (10.4)$$

where:

    ADT = apparent dead time (min)

The two points for each process variable response curve in Figure 10.2 are listed in Table 10.3.

Using Equations 10.2 and 10.3 for the case with $T_1 = 22$ and $T_2 = 0$, the calculated first-order time constant is TFO = 1.5(23.7

**Table 10.3   Data Points for Smith Calculations from Figure 10.2 with Controller Output Step from 50% to 60%**

| DT | $T_1$ | $T_2$ | Time (min) | Process Variable (%) |
|----|-------|-------|------------|----------------------|
| 2  | 22    | 0     | 9.1        | 52.83                |
| 2  | 22    | 0     | 23.7       | 56.32                |
| 2  | 20    | 2     | 10.3       | 52.83                |
| 2  | 20    | 2     | 23.6       | 56.32                |

– 9.1) = 21.9 min, and the calculated apparent dead time is ADT= 23.7 – 21.9 = 1.8 min.

For the case with $T_1$ = 20 and $T_2$ = 2 the calculated TFO = 1.5(23.6 – 10.3) = 20.0 min, and the calculated ADT = 23.6 – 20.0 = 3.6 min.

Many different tuning rules have been published for estimating the controller tuning parameters to use for a first-order plus dead time (FOPDT) process response, and they are too numerous for a complete review in this book. Only a few will be presented.

## 10.4 Quarter Decay Ratio (QDR) Tuning Rules

Corripio[2] presented a summary of the quarter decay ratio (QDR) equations for various controller actions as shown in Table 10.4 for self-regulating process variable responses. For the case with DT = 2, $T_1$ = 20, and $T_2$ = 2, the two-point method calculations gave TFO = 20.0 min and DT = 3.6 min. The QDR equations give a calculated controller gain of $K_c$ = (0.9)(20)/(1.0)(3.6) = 5.0 and $T_i$ = 12.0 min for a PI controller. These recommended tuning constants are a little more aggressive than the optimum tuning constants of $K_c$ = 4.08 and $T_i$ = 13 that were found by pattern recognition in Chapter 9, Section 9.5, for the same system.

**Table 10.4   Equations for QDR Tuning Rules**

| Control Action | Controller Gain $(K_c)$ | Integral Time Constant $(T_i)$ | Derivative Time $(T_d)$ |
|---|---|---|---|
| P | 1.0 TFO/$(K_p$ ADT) | | |
| PI | 0.9 TFO/$(K_p$ ADT) | 3.33 ADT | |
| PID, series | 1.2 TFO/$(K_p$ ADT) | 2.00 ADT | 0.5 ADT |
| PID, parallel | 1.5 TFO/$(K_p$ ADT) | 2.50 ADT | 0.4 ADT |

**Table 10.5   Equations for IMC Tuning Rules**

| Control Action | Controller Gain $(K_c)$ | Integral Time Constant $(T_i)$ |
|---|---|---|
| PI, sluggish | (TFO/$K_p$)/(DT+TFO/2) | TFO |
| PI, moderate | (TFO/$K_p$)/(DT+TFO/3) | TFO |
| PI, aggressive | (TFO/$K_p$)/(DT+TFO/4) | TFO |

# 10.5 Internal Model Control (IMC) Tuning Rules

A set of Internal Model Control (IMC) tuning rules were established by Rivera, Morari, and Skogestad[3] for a first-order plus dead time (FOPDT) open-loop process response that simply involves the adjustment of the proportional gain in the controller, $K_c$, for tuning. The integral time constant, $T_i$, is set equal to the first-order time constant, TFO, for PI controllers (Table 10.5).

The recommended tuning constants using IMC tuning rules for the case described earlier are shown in Table 10.6. All of the combinations of tuning constants are more sluggish than the optimum tuning constants of $K_c$ = 4.08 and $T_i$ = 13 min for excellent response to load disturbance, as shown by the pattern recognition method in Chapter 9, Section 9.5. The minimum IAE (integral of absolute error) from a setpoint change was achieved with $K_c$ = 4.08 and $T_i$ = 22 min as shown in Chapter 9, Figure 9.1.

**Table 10.6   PI Tuning Constants Using IMC Tuning Rules with $K_p$ = 1.0, TFO = 20, and ADT = 3.6**

| Control Action | Controller Gain $(K_c)$ | Integral Time $(T_i)$ |
|---|---|---|
| PI, sluggish | 1.47 | 20 min |
| PI, moderate | 1.94 | 20 min |
| PI, aggressive | 2.32 | 20 min |

Skogestad[4] reported a modification of the IMC rules to SIMC rules for improved disturbance rejection. The recommendation for PI control is to adjust the integral time constant to the smaller value of either $T_i = \text{TFO}$ or $T_i = 8\ \text{ADT}$. The TFO of 20 min is smaller than the calculated $T_i = (8)(3.6) = 28.8$ min in this case.

## 10.6 Integrating Process Response

The process flow diagram in Figure 10.3 shows the concept of an open-loop process response for an integrator. The level in the tank can be envisioned as the integrator or accumulator of the difference in flow rate between the process fluid flow rate into the tank and the flow rate out of the tank. There may be little dead time, if any, between the time the valve position changes and the level change starts to integrate the difference in flow rates.

Another example of an integrating process response is the storage of an inert gas in a pressure vessel. The inert gas can accumulate or can be depleted from a storage tank as indicated by the gas pressure in the vessel.

An integrating process response, for example, the level in a reflux drum or distillation column base, will eventually overflow or run empty in an open-loop step test without further manual intervention. One example of an open-loop step test

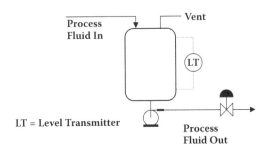

**Figure 10.3   Process concept for modeling an integrator.**

can be with a vertical reflux drum with the level transmitter at 50% at steady-state conditions. Say the controller output is at 70% with the flow rate into the tank equal to the flow rate out of the tank at 0.05 tank volumes per minute. A tank volume can be defined as the volume in the tank between the levels of 0% and 100%. This would give a residence time of 20 min if the tank were at 100% level and 10 min if the tank were at 50% level. Increasing the flow rate out of the vessel from 0.05 to 0.06 tank volumes per minute would cause the tank level to drop 1% per minute. With a constant liquid flow rate into the tank, a step change to increase the outlet valve position would give a liquid level response that would be a straight line sloping down with time.

Some guidelines for level control are to use a proportional gain of 1.25 (proportional band of 80%) in the controller when using the accumulator for level averaging to smooth the outlet flow rate and allowing the level to deviate from setpoint to accomplish this. The integral time constant can be set between 1 and 2 times the residence time of the material in the accumulator. A higher proportional gain can be used in the controller for tighter level control.

All of the tuning rules in Tables 10.2, 10.4, and 10.5 predict that the controller gain, $K_c$, can be set to a high value if the dead time, DT, is short. As a result of this, the proportional gain for a level controller can often be adjusted to very high values without creating a sustained cycle if the integral time constant is kept larger than the closed-loop response time for the process variable to move 95% of a step change in setpoint as shown in Chapter 9, Section 9.9. At some point, an increase in proportional gain in the controller, $K_c$, will simply amplify the noise in the process variable signal into chatter (rapid movement) of the controller output and the control valve position. The noise in the process variable signal can be caused from splashing in a reflux drum or boiling in a reboiler. In that case, the controller gain can be reduced until the rapid movement in the control valve is reduced to 2% of the average

valve position or less. For example, if the average valve position is moving up and down erratically between 58% and
60% with a proportional gain of 3, then the gain could be cut
in half to 1.5 to reduce the band of valve movement to run
between 58.5% and 59.5%. Also, the liquid level signal could
be filtered to reduce the noise level.

An open-loop step test on a level controller would start
with the liquid level and controller output at steady-state values with a constant flow rate of liquid into the vessel. Then,
increase the controller output by a step change, and observe
the rate of change in the liquid level. For example, with the
initial liquid level at 50% and the controller output at 70%,
put the controller in manual output mode and increase the
controller output to 80%. Observe the rate of change of the
process variable, say, a decline of 1.0%/min. This can be used
as the maximum process response rate, MPRR, for calculating
a controller gain, $K_c$, by the Z-N equation in Table 10.2. $K_c$ =
$(0.9)(10)(1)(1) = 9$ for an apparent dead time, ADT, of 1 min or
$K_c = 18$ for an ADT of 0.5 min. This would be expected to give
very tight level control compared to using a gain of $K_c = 1.25$
for level averaging described earlier in this chapter and also in
Chapter 9, Section 9.8.

## 10.7 Summary

This chapter described the characterization of a process
response by introducing a step change in the controller output with a controller in manual output mode. The values of
apparent dead time, first-order time constant, and process gain
were used to characterize an open-loop self-regulating process
response. The values of apparent dead time and percentage
change in accumulator level per minute were used to characterize the open-loop integrating process response. Calculations
for controller PID settings were provided by several different
tuning rules.

After the recommended controller tuning constants have been calculated, they need to be put into the controller and then tested in closed loop for a final evaluation. Closed-loop performance results, that is, with the controller in automatic output mode, can be evaluated by the pattern recognition methods presented in Chapter 9.

## Exercises

10.1 A controller was put in manual mode, and the controller output was changed from 50% to 60%. Two points on the response rate were 51% at 4.1 min and 52% at 6.7 min. What proportional gain and integral time constant would be recommended from the Ziegler–Nichols open-loop response curve method for a PI controller? Would these tuning constants be expected to give a more aggressive controller response than $K_c = 4.08$ and $T_i = 13$ min or a more sluggish response?

10.2 Using the data from Table 10.3 for DT = 2, $T_1 = 22$, and $T_2 = 0$, what proportional gain and integral time constant would be recommended for a PI controller using the quarter decay ratio tuning rules?

10.3 Using the data from Table 10.3 for DT = 2, $T_1 = 22$, and $T_2 = 0$, what proportional gain and integral time constant would be recommended for a PI controller using the aggressive SIMC tuning rules? Would the tuning constants from the SIMC tuning rules be expected to give a more aggressive response than $K_c = 4.08$ and $T_i = 13$ min or a more sluggish response?

## References

1. Ziegler, J. G., and Nichols, N. B., Optimum settings for automatic controllers, *Trans. ASME*, 64, 759, November 1942.

2. Corripio, A. B., *Tuning of Industrial Control Systems*, 2nd ed., ISA, Research Triangle Park, North Carolina, 2001.
3. Rivera, D. E., Morari, M., and Skogestad, S., Internal model control. 4. PID controller design, *Ind. Eng. Chem. Proc. Des. Dev.*, 25, 252–265, 1986.
4. Skogestad, S., Simple analytic rules for model reduction and PID controller tuning, *J. Process Control,* 13, 291–309, 2003.

# Appendix: Solutions to Chapter Exercises

## Exercise 2.1

$R/D = 4$
$D/F = 0.5$
$R/F = (4)(0.5) = 2.$

## Exercise 2.2

$R/D = 3$
$R/O = 3/4 = 0.75.$

## Exercise 3.1

Equation 3.4, key impurity separation power = $10,000/(1)(2)$ = 5,000.

## Exercise 3.2

Equation 3.3, $N_{min}$ = Ln $[(99.5/0.5)/(1/99)]$/Ln 1.12 = 87.3.

# Exercise 3.3

Economical design is 2 (17) = 34 theoretical stages.

# Exercise 4.1

### Table for Exercise 4.1

| Stage | y |
|:-----:|:-----:|
| 4 | 49.30 |
| 3 | 36.12 |
| 2 | 24.74 |
| 1 | 16.04 |

*Note:* Delta y between top two stages = 49.30 − 36.12 = 13.18 mol% C6 change in vapor phase.

Average delta y above and below Stage 3 = (49.30 − 24.74)/2 = 12.28 mol% C6 change in vapor phase.

Average delta y above and below Stage 2 = (36.12 − 16.04)/2 = 10.04 mol% C6 change in vapor phase.

Delta y between bottom two stages = 24.74 − 16.04 = 8.7 mol% C6 change in vapor phase.

The MRT point would most likely be Stage 4.

# Exercise 4.2

Equation 4.3, distillate/feed = (0.500 −0 .001)/(0.999 − 0.001) = 0.498

# Exercise 4.3

Since the feed contains 60% light key component, the vapor above the feed tray would be expected to be above 60%. This would give 50% light key component below the feed stage, so the MRT would be expected to be below the feed stage.

# Exercise 4.4

The vapor composition above the feed stage would be expected to be above 3% light key component, so the feed stage would be the closest to 50% in the vapor phase. All of the other stages would be expected to be lower in concentration of light key component.

# Exercise 5.1

A scheme that uses a temperature controller to manipulate the steam flow rate to the reboiler. There may not even be any reflux flow to the column. The top vapor temperature may be the MRT (most responsive temperature) point.

# Exercise 5.2

A scheme that uses a constant steam flow rate to the reboiler pot and temperature controller that manipulates the distillate flow rate. The reboiler pot temperature may change the most, but the pot temperature would be an indicator of the distillate composition.

## Exercise 5.3

Either a scheme using a fixed reflux flow rate and an MRT controller that manipulates the steam flow rate to the reboiler, or a scheme using a steady steam flow rate to the reboiler and an MRT controller that manipulates the reflux flow rate.

## Exercise 6.1

The maximum constrained value is 20,000 lb/h reflux because it is 100% of the flow sensor output.

## Exercise 6.2

Using Equation 6.1, $S = (0.02) (39000) = 780$
Equation 6.2, NTS = $\text{Ln}[(1000/2)(1 - 1/780) + 1/780]/\text{Ln } 780$
= 0.93 theoretical stages.

## Exercise 6.3

Using Equation 6.1, $S = (0.02) (39000) = 780$
Equation 6.3, $N_{OL} = \{\text{Ln}[(1000/2)(1 - 1/780) + 1/780]\}/\{1 - 1/780\} = 6.2$ liquid phase mass transfer units.

## Exercise 6.4

Using Equation 6.9, $N_{OG} = \text{Ln}(200/0.1) = 7.6$ gas phase transfer units.

# Exercise 6.5

Using Equation 6.6, $N_{OL}$ = (1) (Ln 3)/(1 − 1/3) = 1.65 liquid phase mass transfer units.

# Exercise 6.6

Using Equation 6.1, $S$ = (1/350) (1.1/0.1) = 0.0314
and Equation 6.8, $N_{OG}$ = {Ln[(5000/1)(1 − 0.0314) + 0.0314]}/ {1 − 0.0314} = 8.76 gas phase mass transfer units.

# Exercise 7.1

DNS = 0
Failure rate = ½
(½)(21) = 10 or 11 samples per week would be out of spec.

# Exercise 7.2

Failure rate of 1/790 would require DNS = 3.0 sigma
0.3% − (3.0)(.025) = 0.3 − 0.075 = 0.225% ethanol

# Exercise 7.3

The MRT setpoint should be raised to increase toluene concentration and reduce the benzene concentration left in the bottoms.

## Exercise 7.4

The separation power base could be reduced to reduce energy consumption, and let the impurity concentrations increase in the distillate and bottoms.

## Exercise 7.5

Rule No. 5 would fire.

## Exercise 7.6

Lowered.

## Exercise 7.7

$0.3(5) = 1.5°C$.

## Exercise 8.1

CO = change in PV times $K_c$ = 5% (5) = 25% change in controller output.

## Exercise 8.2

For Equation 8.1,
$K_c$ = 12.5 and
CO − $CO_b$ = 7% change in steam valve position.
So, $e$ = 7/12.5 = 0.56% of temperature span
And $(0.56/100)(200°C) = 1.12°C$ offset
The new steady-state temperature would be too low.

## Exercise 8.3

The controller gain, $K_c = 100/25 = 4$.
The time to change the controller output as much as the
  gain would change the controller outlet is $T_i = 1/$ repeats/
  minute $= 1/0.05 = 20$ min.

## Exercise 9.1

The integral time constant of 3 min is 13 times smaller than
  the response time of 39 min. Increase the proportional
  gain from 1 to 13, and increase the integral time constant
  from 3 to 39 min.

## Exercise 9.2

The integral time constant should be reduced to about 7
  min, and the proportional gain should be reduced to
  about 1.2.

## Exercise 9.3

Use a smaller setpoint change, so that the controller output
  does not saturate at 100%. Repeat the test with, say, a 5%
  setpoint change in the opposite direction.

## Exercise 10.1

$(52 - 51)/(6.7 - 4.1) = 1/2.6 = 0.385\%$ PV change/minute
Apparent DT $= 4.1 - 2.6 = 1.5$ min
$K_p = 1.0$
The Z-N recommended $K_c = (0.9)(10)/(0.385)(1.5) = 15.6$

$T_i = (3.33)(1.5) = 5.0$ min

These tuning constants would be expected to be much more aggressive than $K_c = 4.08$ and $T_i = 13$ min.

## Exercise 10.2

TFO = 21.9 min

Apparent DT = 1.8 min

$K_p = 1.0$

The QDR recommended $K_c = (0.9)(21.9)/(1.0)(1.8) = 11.0$

$T_i = (3.33)(1.8) = 6.0$ min

The high gain of 11 is much more aggressive than 4.08, and the $T_i = 6.0$ min is much more aggressive than 13 min. The control loop would probably run with sustained cycling.

## Exercise 10.3

TFO = 21.9 min

Apparent DT = 1.8 min

$K_p = 1.0$

The IMC recommended $K_c = (21.9/1.0)/(1.8+21.9/2) = 1.72$

The low gain, $K_c$, of 1.72 is much more sluggish than $K_c = 4.08$

The IMC recommended $T_i = 21.9$ min

The SIMC recommended $T_i = (8)(1.8) = 14.4$ min

The lower $T_i$ of 14.4 min is nearly the same at 13 min so, on balance, the SIMC calculated values would be expected to be more sluggish than $K_c = 4.08$ and $T_i = 13$ min.

# Index